D1011201

The INNOCENT ASSASSINS

Reconstruction of *Homotherium serum*, Zoological Museum, Helsinki University.

BJÖRN KURTÉN

The INNOCENT ASSASSINS

Biological Essays on Life in the Present and Distant Past

Illustrations by Viking Nyström

Translated by

BJÖRN KURTÉN & ERIK J. FRIIS

Columbia University Press • *New York*

Library of Congress Cataloging-in-Publication Data

Kurtén, Björn.
[Skuldlöse mördarna. English]
The innocent assassins : biological essays on life in the present and distant past / Björn Kurtén ; illustrations by Viking Nyström ; translated by Björn Kurtén & Erik J. Friis.
p. cm.
Translation of: De skuldlöse mördarna.
Includes index.
ISBN 0-231-07276-7 (durable acid-free paper)
1. Evolution—Popular works. 2. Biology—Popular works. 3. Paleontology—Popular works. I. Title.
QH367.K9713 1991
575—dc20 *90-43447*
CIP

Columbia University Press
New York Oxford

Book design by Mary M. Ahern

Casebound editions of Columbia University Press books are Smyth-sewn and printed on permanent and durable acid-free paper

Printed in the United States of America
c 10 9 8 7 6 5 4 3 2 1

C O N T E N T S

The INNOCENT ASSASSINS

CHAPTER ONE

Introduction—Or, Was Dalton Wrong?

Nobody says Dalton was wrong.

John Dalton was a British scholar who nearly two centuries ago formulated the *atomic theory of chemistry,* which, among other things, states that elements consist of atoms that have an element-specific mass. Apparently, this theory is still upheld, even in the most enlightened circles; at least, I cannot remember reading triumphant statements like "Dalton was wrong! The atom theory of chemistry definitively rejected! If Dalton were alive today, he would not be a Daltonist!"

Neither can I recall seeing any spectacular public rejections of Einstein's famous equation of the relationship between energy and mass, $E = mc^2$, as being totally out of date and incorrect, which would seem to suggest that it remains generally accepted by people at large as well as by physicists.

These theories remain cornerstone within chemistry and physics, respectively, and it is nice to see that they are so

generally accepted. I fear, however, that their scientific importance is not the real reason. Wisdom cometh by suffering. Dalton and (in this case) Einstein simply are not lawful prey.

As regards Dalton, the case is simple enough. Very few people know his name and still fewer his work. There is no sensation to create and no money to make by "unmasking" him as an out-of-date humbug. In the case of Einstein's equation, other considerations apply. Einstein is, in fact, lawful prey, and many other theories by Einstein have been rejected, often with big headlines (and in some cases deservedly so). This particular equation, however, is a hopeless case. It forms the basis for nuclear power and the atom bomb, and it is difficult today to diddle people into believing that nuclear power and atom bombs are myths.

Unfortunately, the same cannot be said about the cornerstone of modern biology, the theory of evolution in its Darwinian formulation. More or less regularly, there will arise, for instance an author, a journalist, or a scientist (rarely a biologist) to make an authoritative utterance: Darwin was wrong; Darwin has now been definitively rejected; Darwin was okay but he didn't know what we know today; Darwin would not, today, have been a Darwinian; and so on. The peculiar thing is that these *pronunciamientos* are never uttered by serious students of evolution, or affect them in any way. They continue serenely to follow Darwin for the simple reason that his explanation is the only one that can explain evolution, and does it completely—like the chemists, who obstinately go on believing in atoms. This fact, however, tends to remain obscure, as far as popular knowledge is concerned, be-

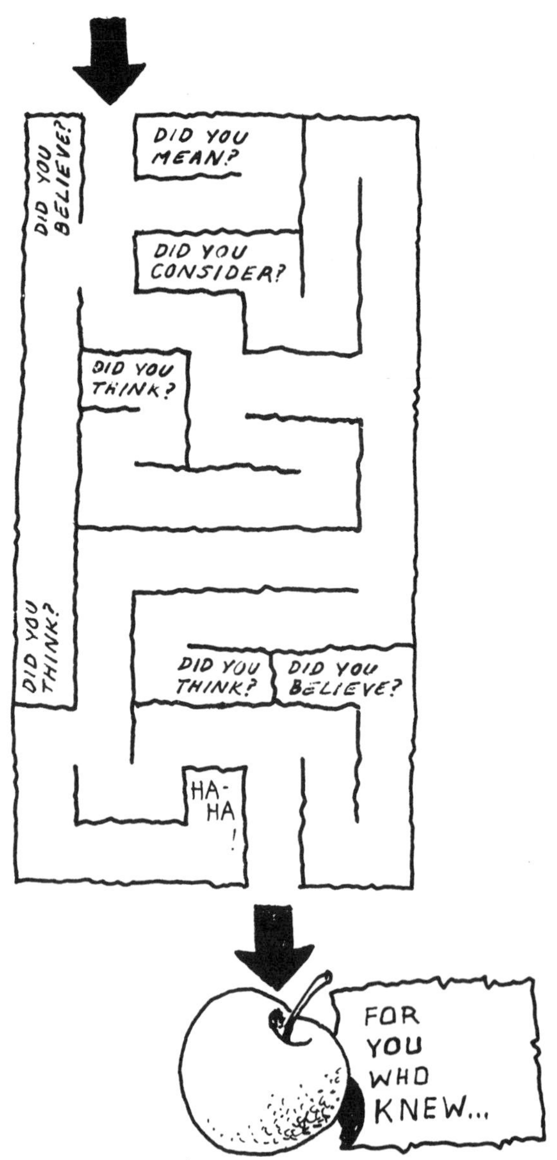

DID YOU BELIEVE?
DID YOU MEAN?
DID YOU CONSIDER?
DID YOU THINK?
DID YOU THINK?
DID YOU THINK?
DID YOU BELIEVE?
HA-HA!
FOR YOU WHO KNEW...

cause the vociferous know-it-alls are the ones who are heard, and dethroning Darwin remains a sensation, although you might think it should be rather threadbare by now. The biologists themselves do not as a rule say much, and they have good reason: biological facts cannot be explained in simple catchwords, and a reasonably explanatory answer necessitates a fairly long article. Which the media do not publish, because there is no room for it. And so the public remains unenlightened. Of course, many biologists write excellent books for the general reader—names like Futuyma, Richard Dawkins, J. B. S. Haldane, P. B. Medawar, and others come to mind—but they only reach a few.

Biologists working on evolutionary problems, then, find that they have an unexpected extra job on their hands, which most other scientists escape: they have to waste their time explaining and defending their work in the

marketplace (to quote John Maynard Smith). However (he goes on), it is useful to us biologists. Nothing clears the brain better than having to explain your job to persons who are not specialists in it. Incidentally, it turns out that we can turn to Darwin himself in many cases: he very carefully considered all the evidence against his theory that he himself or his contemporaries could think of and wrote conscientious answers. Much of what is said against Darwin today is a hodgepodge of the same Victorian arguments, which have been dusted off and served up innumerable times since the year 1859.

Why have many highly intelligent and learned intellectuals been so hostile to Darwinism? Probably in part because they disliked thinking of themselves as products of blind chance (which they falsely thought that evolutionary theory implied) and in part because they saw natural selection (evidently in the distorted shape given it by Spencer) as brutal and inhuman. They were wrong on both points. No modern evolutionist imagines that life originated and evolved by chance, and Darwinism (the theory of natural selection) is something very different from the principle of "nature red in fang and claw." Shaw, Koestler, and others have been led astray by their own caricature of biology.

This book was not written as a "defense" of evolutionary theory. However, I try to give an idea of what it means to look at the world from the point of view of biology, especially evolutionary biology. I suspect that much of what is said will be unexpected and surprising to those readers who are not familiar with this useful exercise. Biology is probably one of the most misunderstood sciences in present-day society. This is absurd and dangerous. No other science advances so rapidly and on such a broad front as biology—including special fields like medicine, forestry, and so on: they all belong together. No other science has equal importance for all of us, here and now, to say nothing of the future.

My topic is not applied biology and biotechnology, which are the fields where the work is done that immediately influences our existence. To get them in perspective it is necessary to go to the basic principles, and they are the topic of the introductory essays. Since my own perspective is that of paleontology, the history of life and its evolution, my examples are taken mostly from the past.

Much of science is done simply for money or credit.

The result is rarely inspired, generally reliable, but in the worst cases dishonest. "We are the guardians of the integrity of Science," says Jacob Bronowski, meaning intellectual as well as moral integrity. It is best maintained by what is the basis of all good science: the curiosity about *what things are really like.* Pseudoscience, on the other hand, is produced by those who "know" beforehand what the answer is going to be. Some time ago I read the following headline: "Professor Cousteau starts a new expedition to Santorin to prove that this island is ancient Atlantis." That headline, I am sure, was unfair to Cousteau; I do not think he went out to *prove* some preconceived opinion, but just to find out what things were like. Of course, no kind of research is carried out without some idea as to what you may find out; that is quite all right, but you have to be ready to change your opinion. Changing one's opinion is very stimulating. The politician who vehemently denied the accusation that he changed his opinion, like his shirt, once a month, did not know what he was missing. A piece of good advice for everybody is to change your opinion in some matter, big or small, every day. It keeps the soul young and opens up new horizons. But the reason for the change must be valid.

Pseudoscience, of course, has great popular appeal, as a way to cut corners in the search for happiness. The Committee for the Scientific Investigation of Claims for the Paranormal (CSICOP) has an annual award (a bent spoon) that is presented to the silliest pseudoscientific project of the year. In 1981, it went to the Pentagon for a six-million-dollar investigation of whether Soviet missiles can be destroyed by burning photographs of them. It may be vaguely reassuring to some of us to realize that the Pentagon, like so many other military establishments, is being run by quite normal fools.

Essentia non sunt multiplicanda præter necessitatem, the simplest explanation is the best; this thesis was formulated more than six centuries ago by the Franciscan monk who was called *Venerabilis inceptor,* the Venerable Pioneer. Under the name Ockham's Razor, or the principle of parsimony, it still works as an eminent rule of thumb in all empirical research. In biology, however, the relationships are often so complicated that even the simplest possible explanation becomes hard to survey and may lead one to think of the classic German travesty of Ockham, *warum etwas einfach machen, wenn es auch kompliziert geht* (why make something simple, when it's just as easy to make it difficult)? The principle of, say, natural selection is simple enough, but an exhaustive description of how it works in nature would fill many volumes.

CHAPTER TWO

Biology and the Bookshelf

It is interesting to speculate on the image of science as it appears in fiction—especially science fiction. You will soon realize that, although our lives today are utterly dependent on science, those who really try to find out what science is all about and what it has to tell are frighteningly few. Superstition and pseudoscience have far greater appeal. They hold out the prospect of shortcuts to happiness, while science forces us to face the unpleasant truth; and so we turn to astrologers to foretell a brilliant future, to dowsers to prospect for water, and to soothsayers to find missing persons, although we should know that most things have to be done the hard way.

It may be that scientists, in the popular view, rank simply as one kind of wizard. Science can do seemingly miraculous things, like picturing at close range the rings of Saturn, or tell us what our ancestors looked like 400 million years ago. Wizardry—well, why not then accept any kind of occult mumbo-jumbo? Or, as an alternative, regard science with plain distrust. Skepticism is a sound reaction, but all too often is carried to absurdity. Science fiction abounds with mad professors who seek world domination—vulgar versions of Faust.

In part this may be due to confusion between science and arms technology—a common error. But the roots go deeper. As Ernst Mayr has pointed out, our idea of the universe is mainly based on eighteenth-century thought, the Age of Enlightenment; and that, in turn, was founded on physics, chemistry, and astronomy. Now, in their present-day form, these branches of science are presenting us with a view of the universe that is frightening in its immensity, emptiness, and monotony—an endless space with occasional blobs of matter, floating like specks of dust in the void. Aha, we tell ourselves, that is the world of science, and promptly turn our backs on it, realizing quite properly that a single human being is greater than a thousand suns.

But the world of physics is no more than a corner of the world of science. A human being is greater than a star.

A bacterium is greater than a star. The star cannot become very much else than a star (oh yes, it can explode and become a black hole, so what?)—but bacteria have evolved into people. A proton is like all other protons, they are interchangeable. A bacterium is unique and, moreover, is changing from moment to moment. Biology is the study of individuals, all of them unique, in inexpressible multiplicity and richness. And so the world of biology is something very different from the world of physics, and calls for a different approach and different concepts (which I will discuss further in a later chapter). And, since we ourselves are living beings, the place of biology is in the center of that sequence which leads from the humanities on one flank to physics on the other. C. P. Snow was a physicist and an author but not a biologist; hence his "two cultures" of science and the humanities.

Few indeed are the science fiction writers who have realized this. It is symptomatic that SF usually deals with technology, astronomy, and physics—spaceships and robots, hocus-pocus with time and distance. (I do not overlook the fact that much of the best SF is social criticism; I speak only of its paraphernalia.) Of course there are brilliant exceptions like Olaf Stapledon, the incomparable grand master of the genre. Yet I hold that biological thinking would do much for SF and serious writing in general (John Gardner, for one, appreciated this). And this is because facts in this field are so wonderful that most comparable products of imagination tend to fall flat. I am now going to present some instances of biological piffle (there will be others later on). I do it not only to be malicious (although that can be fun) but to point to other possibilities. My examples will run the gamut from the trivial to important principles.

To begin with the trivial: it happened many years ago that a certain traveler, Charles Miller, encountered a living dinosaur in New Guinea. (Unfortunately, as is the rule in these cases, his camera didn't work properly.) This could have developed into a magnificent, thumping lie, but unfortunately Mr. Miller suffered from complete biological innocence and so described a trite nonfunctional cross between various common or garden variety dinosaurs. (What can really be done in dreaming up new forms of life is shown by Dougal Dixon in his wonderful book *After Man*—examining what the animal kingdom of the future will look like when humanity, to the great benefit of most other life forms, has destroyed itself.)

There is a popular misconception that you can find to satiation in literature, beginning with Genesis. It is the concept of a "racial memory," dear to the heart of many a writer. We are told that the images and ideas experienced by our distant forebears are still mysteriously on tap in our innermost being. This idea is ancient indeed, being

voiced for the first time in the story of Jacob tricking Laban by breeding cattle that is "ringstraked, speckled, and spotted" by means of methods Lamarckian, or more properly Lysenkoan. All this is of course just as fatuous as blaming your own silliness on your mother being frightened by a donkey. From genes to body and mind is a one-way street; there's no traffic in the other direction.

Yet we probably have "racial memories" in a very different sense. They are patterns of behavior that were fixed the hard way, by way of mutation and natural selection. That may be the reason why some of us are unreasonably terrified by snakes or spiders, or why a woman holds her baby so that it will experience the great rhythm of her heartbeat. These are behavior patterns that have had, or still do have, their use in the preservation of the species. Very likely the incest taboo arose in the same way. It is striking that children (regardless of relationship) who grow up together very rarely develop serious mutual sexual attraction. Intuitively the great poets have seen this, for in literature incestuous relationships, from Oedipus on, are enacted between people who do not recognize each other from the past.

In our anthropocentrism we deny that our biological nature can affect our behavior. Typical is the huge outcry against the first (and naturally, very preliminary) attempts to analyze human behavior in terms of sociobiology. (The protest is in fact an excellent example of territorial, i.e., biological, defense against an intrusive neighbor science.) And yet it is in the area of intersection between sciences (in this case sociology and biology) that some really exciting discoveries are made. Instead of constructive criticism, however, we have heard diatribes on the political naïveté of biologists. Rarely, as Svante Folin has pointed out, do we hear any laments about the biological naïveté of politicians, a much more common and more serious problem.

Another false notion is that biologists, like dog breeders, are concerned about "purity of race." Probably this is because the Nazi self-styled "race biologists" had the gall to refer to themselves as "biologists" when they were only practicing the most degraded pseudoscience. In fact nothing could be more wrong. To a biologist it is clear

that variation, multiplicity, is the essence of life; and this is true for humanity as well. Our greatest treasure is our variety. (The cloning of invariant human genotypes is an abiological nightmare, to be entertained only by the totalitarian mind.) To Andor Thoma, we are all of us leaves on the great tree of humankind, and none should be lost. All of us who are now living are brothers and sisters, says Stephen Jay Gould; this is not political rhetoric but sound biology.

Most of the essays in this book deal with my own corner of science, paleontology, the study of the life of the past. We meet the fossils in museums: bones, teeth, jaws, empty eye sockets. Some of them seem larger than life. They may appear comical, awesome, or at best heroic, sometimes with a touch of abstract beauty. Life is gone: gone, the softness of the fur; gone, the color pattern of the plumage; gone, the flow of motion; gone, the spark of life in the eye. And so the past world, to the museum visitor, may seem like a nightmare of creaking bony frames, menacing eyeteeth, sharp horns—many a writer has walked into that trap.

Take the great flying dragon, the pterosaur or flying reptile. Old reconstructions show it dark and dreadful with its immense snapping beak and black leathery wings. Conan Doyle, in his *Lost World,* makes it even more hideous by steeping it in the stench of carrion. And yet, as Adrian Desmond has shown *(The Hot-Blooded Dinosaurs),* a great flying reptile, seen in life, would probably have impressed us as a vision of exquisite beauty, soaring on white wings like an albatross over a crystal sea.

There it is in a nutshell. Out of the past, people and animals face us in their *pompa mortis;* and so we are beguiled into figuring the ancient world as something horrible, populated by sundry monsters. It is like frightening a child with a paper skeleton.

Great beauty can be found in the past. What is beauty? Why do we see something as beautiful? And what does beauty have to do with evolution?

CHAPTER THREE

Tom Thumb and King Kong

The motion picture *The Incredible Shrinking Man* was released in 1956. I don't think many remember it now. It was based on a novel by Richard Matheson and directed by Jack Arnold, and it represented a new, interesting variant on an old theme. The hero, Scott Carey (played by Grant William), shares the fate of many a fairy-tale figure, miniaturization; but with a difference, for he is not dwarfed by a single stroke of magic but shrinks gradually. That, by the way, is the difference between the world of the fairy tale and the world of science fiction. For some time Carey lives in a doll's house, but he continues to shrink and eventually is lost among the atoms. The mood of the film is serious and, at the end, evokes a cathartic transfiguration.

While on his *via dolorosa,* Carey encounters terrible dangers: an ordinary house cat becomes a supertiger; later on, when he has shrunk to insect size, he meets with a monstrous spider.

When Gulliver comes to Lilliput, he too finds miniature people. Tom Thumbs, imps, and hobgoblins are popular fairy-tale figures. A biologist, of course, will promptly point out that such beings with human traits are unthinkable. Assume a human being reduced to one-hundredth of the normal length—that is, 18 millimeters tall (i.e., three-quarters of an inch). However, a living cell is not viable unless it reaches a certain size. Thus the number of cells in the body would be reduced to a small fraction. (True, small mammals tend to have somewhat smaller cells than large ones, but the difference is not very great.)

Take the eyes. The area of the retina would be reduced to one-hundredth of a hundredth, and so contain only 1/10,000th the number of cells in our own retina. The vision would become coarsely granular, like a greatly enlarged photograph, and our miniature person would have very bad eyesight. As to the brain, that would be even worse. Its volume would go down by one hundred times one hundred times one hundred, to one-millionth of our brain. Such a brain might still be able to control the main bodily functions but certainly nothing more; so our miniature human being would be abysmally stupid.

Still, in the world of science fiction nothing is impossible. Why not, then, let Carey shrink and yet retain his humanity—we might, for instance, let all atoms, molecules, and cells in his body shrink at the same rate. What would the world look like, to him?

In the film, it is painted in dark colors: the house cat becomes a voracious tiger, and the encounter with the spider is as horrific as a meeting with a tyrannosaur. In spite of his small size and weakness, however, Man triumphs: his weapons are his intelligence and a pin for a lance. All this is very realistic and seemingly to the point, and we might well believe that the world would have looked just like that, to him.

But it wouldn't, not one bit.

In one sequence, Carey falls down the cellar stairs and almost breaks his neck. Now, in the real world he wouldn't have done that. When his length was reduced to one-hundredth, his weight (like his brain volume) was reduced to one-millionth. He would have weighed one-tenth of a gram and have fallen as lightly as a feather, carried by the resistance of the air.

Also, what about his helplessness? The strength of a muscle changes not as its mass but as its cross-section, i.e., by an area, not a volume. The same holds for the bones in Carey's skeleton. And the area changes (as did the retina) with the square of the length, not with the cube as did brain volume and weight. When Carey's weight was reduced to one-millionth, his strength was only reduced to 1/10,000. Relative to his weight, mini-Carey would be one hundred times as strong as the normal Carey.

The film shows Carey, panting and exhausted, pull himself up a sewing thread that he has anchored to the side of a box, using the pin as a grappling iron. Elsewhere, he is shown at the chink between two floorboards, not really daring to jump. In reality there would have been no problem whatever. He could jump like a flea. And this would have robbed the encounter with the earth-bound spider of most of its terror: with a single jump Carey could escape.

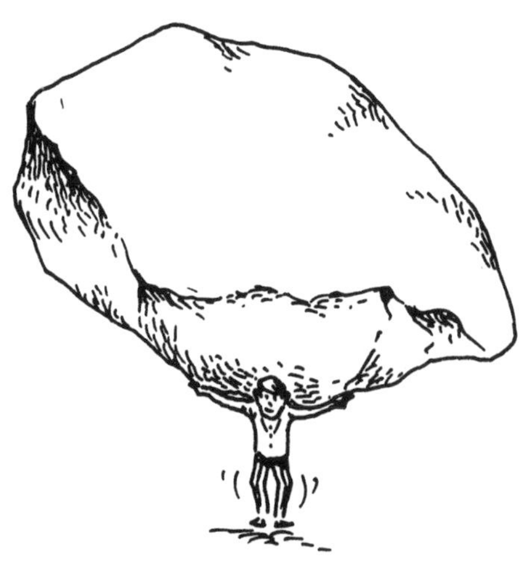

Fleeing before the cat, Carey is seen running like a normal human, with a speed that looks normal to us—and at the same time, we see the quick movements of the cat, frighteningly enlarged. But a mini-Carey would certainly not move in the same manner as a normal human. Our legs function like pendulums, oscillating at a rate that is mainly determined by their length (though it may be modified by extra muscular development). If we want to increase the frequency of the oscillation, that is to say run, we do it by reducing the length of the pendulum: the leg is bent at the knee. When a small child runs, it takes many more steps per second than a grown-up. In a cat, the legs swing at a still faster rate.

For this reason, size is not so important for speed as you might think. A short runner may defeat a tall one; a small dog runs about as fast as a large one. Of course, mini-Carey would not be able to reach the same speed as the normal Carey; for one thing, the wind resistance would be much greater. But he would certainly have a more-than-even chance against the cat.

That does not hold just for running, but equally for other movements, and indeed, for reactions in general. The nerve impulses move at a finite speed; reduce the length of the nerves to one-hundredth, and the time from

sensory impression to action is similarly reduced. (Perhaps I am cheating here. I have let all Carey's atoms shrink—maybe I should let his nerve impulses slow down to the same degree. If so, I ask your pardon.) This means that the time dimension would change completely for the mini-man. Our sense of time depends upon our rate of living. A mouse lives just as much as an elephant, only at a much faster rate—for instance, the hearts of the two animals make about the same number of beats in a lifetime. In the three-quarter-inch format our rate of living would be still faster. A single day would seem as long and as full of experience as several months. The quick movements of the cat, to Carey, would appear as ponderous as those of an elephant to us, and the rattling race of the monster spider would change into the slow crawling of a tortoise. To a spider, in fact, Carey would be an overwhelming adversary, a demon of invincible strength and lightness of movement. Far from being a powerless wretch in this world, Carey would be its undisputed ruler.

There are many other amusing consequences. A light soprano voice would become a thick, unintelligible murmur to him, if indeed it remained audible at all. Instead, he would be able to hear new sounds, much higher than those now audible. The whine of a mosquito would sink six or seven octaves and become something like the roar

of an airplane engine. The sonar signals of the bats would become shattering trumpet calls.

In the movie, Carey is shown swimming, but in the real world he would be able to walk on the water, provided that the surface was smooth—it would be something like walking on a tough, yielding skin of ice. In contrast, it would be difficult for him to get *into* the water, and when there he would find it well nigh impossible to get out again!

It could be worthwhile to make a movie along these lines. The result might be entertaining as well as instructive, but it would also be more in the Superman genre, perhaps lacking the quality of a solemn struggle against the forces of nature that made *The Incredible Shrinking Man* so interesting.

We cannot become Tom Thumbs; but we have all been something less extreme in the same line, to wit, children; and we have all experienced changes in the rate of living during our lives. The older we get, the faster the time seems to flow. The Swedish zoologist Gaston Backman concluded that the time of an organism is *logarithmic*. If we take the logarithmic scale shown here and assume that it represents years, we can see that it demonstrates exactly what every human knows: in childhood the years are long; the older you get, the shorter they become. To a child, one day is an eternity; to the aged, the days pass all too quickly. Experiences and developmental processes that may be accommodated in a few hours for a child require much longer stretches of time—days, weeks—in the life of the grown-up. A child learns a new language in a short time; to the grown-up it means an arduous long-term project. A child makes its decisions quickly, while the old take their time thinking it over. You may believe that this is due to the wisdom and experience of age, but it might also reflect the difference in the tempo of living. Maybe

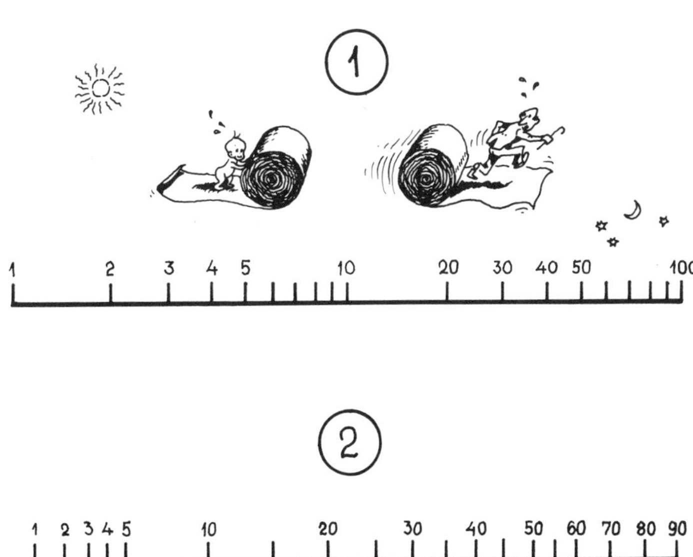

the young person who makes a choice in five minutes thinks it over just as much as the old one who needs an hour. That does not mean that the young person's choice is wiser, for the old have the advantage of longer experience.

This pertains to bodily processes as well. A child's wound heals rapidly, while scar tissue may take a long time to form in the old. The difference in rate of living is reminiscent of the difference between small and large animals; the small animals, with their rapid movements, make a "young" impression, while the large ones seem "old."

Well and good. But am I not beating a dead horse? The day is long to a child, time runs out of the hands of the old—has this not been repeated innumerable times? On the other hand, have you thought about what it actually means? For the sake of argument, look once more at the log scale. The distance from 1 year to 10 is *the same* as that from 10 to 100. Maybe this telescoping is a bit exaggerated. We could make it less extreme by letting 11 represent one year, 12 represent two years, and so on, as in the second log scale, although mathematically that is objectionable. Yet even in this case the distance from 1 year to 15 is almost as long as that from 15 to 50.

We cannot escape this conclusion: *During one-half of our subjective lifetimes, we are children.*

This is a truth worth considering by policymakers who still think that the next best is good enough for kids and who will cut school standards first when it comes to saving money.

During half our lives we are children, with all that implies for our needs: the need to turn toward love, knowledge, generosity, tolerance, creativity, all that long line of cherished goals, and to combat hate, antirationalism, egotism, racism, and that withering of the soul from which so many suffer needlessly because they were never taught anything better. The list may be extended as far as you like. Look, for instance, at books, TV movies, and theater for children, and see what the reality looks like.

Since we cannot become Tom Thumbs—what about giants? Gulliver visits Brobdignag, which is inhabited by giants so colossal that he can ride astride a woman's nipple. King Kong easily embraces the Empire State Building and climbs to its summit. Let us once again consult our cross-grained biologist; he will brand the thing as impossible. A being 100 times as tall as a human being (i.e., 180 meters) would weigh one million times as much as a human being (i.e., 100,000 tons). On the other hand, the strength of this giant would only be 10,000 times as great as ours, or in relation to weight, only one-hundredth of ours. Kong's bones would break like matchsticks, he would collapse into a mush on the ground, and of course die then

and there. (To say nothing of the skyscraper, which would be crushed under his weight.)

The weight problem might be solved by putting Kong on a very small planet, where, however, unfortunately he would suffocate, because a planet with such a small gravitational pull would not have an atmosphere. Another possibility is to keep Kong in the water, a carrying medium in which—as demonstrated by the whales—you may grow very big indeed.

Yet there would still be problems. The fact that the volume grows as the cube of the length, and the surface as the square, will inevitably lead to difficulties—for instance, as regards the lungs. They absorb oxygen through the surface of the pulmonary vesicles, which, in this case, would be expected to feed a hundredfold body mass. Disregarding such difficulties, we may expect a life which would be antithetical to that of Tom Thumb. Our life would be lengthened, but we wouldn't notice that because its tempo would be correspondingly slowed. A deep male voice would be heard as a quacking Donald Duck noise, and an elephant would be seen to move quickly and jerkily like an ant.

King Kong is a biological absurdity. But how about enlarging small creatures like insects and spiders? A not uncommon theme in science fiction is the giant insects that become the masters of the world.

Spiders as well as insects are arthropods, animals that, in contrast with vertebrates, have an outer skeleton. Among water-dwelling arthropods, some have reached surprising dimensions. These are the sea scorpions or eurypterids, which existed for about 200 million years; the last died about 250 million years ago. (Living land scorpions may well be derived from early sea scorpions.) The biggest sea scorpion on record so far reached a length of 1.8 meters,

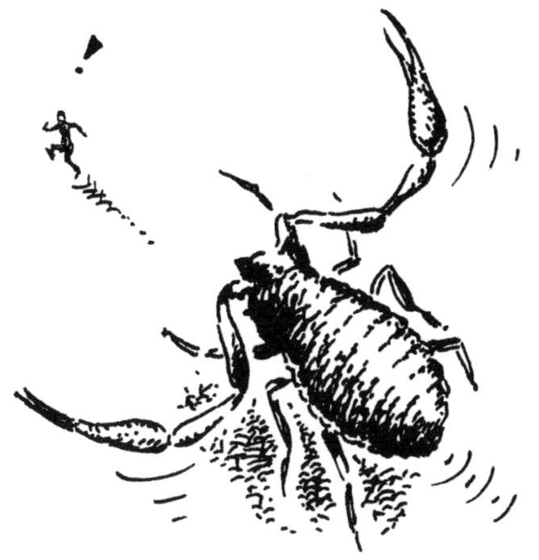

that of a tall man. The living Japanese giant crab has a smaller body but enormously long, spidery legs, giving it a total diameter of up to four meters.

To the insects, which breathe air, it does not appear possible to attain such large sizes. This is due (among other things) to their lack of circulating blood, which in vertebrates, for instance, transports oxygen from the lungs to every part of the body. The insects, instead, have tracheae—numerous inlets from the surface of the body, an arrangement that definitively limits the possibilities for growth. And so it happens that even the biggest insects (a dragonfly from the Carboniferous period had a wingspan of 70 centimeters) are comparatively small animals. In fact, the largest land arthropods are certain land crabs, which do have blood circulation.

Otherwise, too, a giant land arthropod would have problems to face. The lack of an inner skeleton would necessitate the development of an immensely heavy and clumsy armor. A lion-sized spider would probably collapse under its own weight. (We may recall this when reading about the spider-monster Shelob in Tolkien's Ring trilogy.)

These speculations are not as childish as they might seem. In fact, they lead us to a biological theme called "scaling": the problems resulting from changes in size. (Professor Knut Schmidt-Nielsen published a pleasant book on this in 1984, *Scaling: Why Is Animal Size So Important?*)

Even quite modest changes in size will require certain adaptations to ensure an optimal result, and this, of course, is because volume (weight) and surface change with the cube and the square, respectively, of the length. Changes in size can often be traced in the fossil history. (Increase seems to be somewhat more common than decrease, but

both are known to occur.) The living brown bear and grizzly bear *(Ursus arctos),* for instance, arose from a much smaller species that lived before the Ice Age; Ice Age bears were larger than those of the present day, and during the last 10,000 years, body size has decreased again. Wherever there is a full record it can be seen that the changes have been gradual (even when comparatively rapid) and not in leaps—quite naturally, since each change requires other adjustments to result in a well-integrated organism. All such changes occur within a network of interconnections and feedback between anatomy and function.

In the fairy tale, change is instantaneous, like the step from life to death: a human is transformed into Tom Thumb, a toad into a beautiful prince; even Pinocchio's nose grows erratically. In science fiction, change is gradual, yet still on the individual level, like natural things that grow and fade. Both may be charged with insight into our nature, our conditions, our fate. In the biological reality it is the race that changes, at a rate which, seen

geologically, may be rapid or slow, but to the human beholder is so infinitesimal that we take the status quo for granted in our everyday life. And yet the biological perspective too may be of decisive importance to our nature, our conditions, and our fate.

CHAPTER FOUR

Why Should We Feel the Countryside Is Beautiful?

The encounter between the "two cultures" is rarely problem free. That science and the humanities can enrich each other is rarely acknowledged, though it is acknowledged perhaps more often by scientists. Nevertheless, interfaces exist, for instance within anthropology—ideally, an integrated whole ranging from ethnology and archaeology to physical anthropology and primatology. This field bridges the gap between the arts and the life sciences (and also geology). Another interface may be seen in historical geology and paleontology: as the great American paleontologist William D. Matthew put it (borrowing from Milton), "Paleontology is but History writ large."

You don't have to be particularly interested in biology to feel that the countryside is beautiful. ("Countryside" here means any reasonably unspoiled natural environment.) Not so many, perhaps, have wondered *why* this should be so.

To begin with, however, I should like to contrast pure aestheticism with that world view which is based on

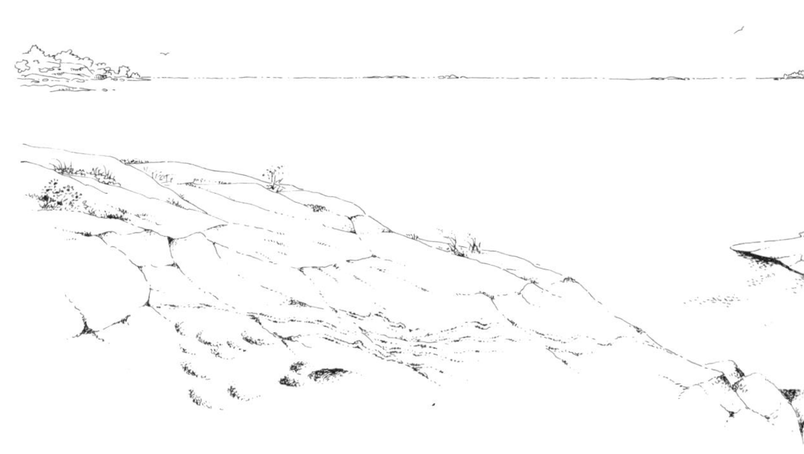

knowledge. My colleague Nils Edelman, a geologist, has commented on the idea that knowledge tends to kill feeling—for instance, that you stop enjoying birdsong when you know why the bird is singing. On the contrary, he says, speaking of the lovely ice-polished rocks in the Baltic archipelago: as a youngster he appreciated them mainly according to their suitability for sunbathing. Now their geological history invests them with a new kind of beauty. I agree. They are not just scenic: they do open a dizzying vista of worlds in collision a billion years ago, followed by a much later interval when the ice gave them their present shape. At the same time, we get a profound sense of the perishability of the landscape—of the fact that it exists only now and to us, and that we should face it with the same tenderness that is due to all things that flower and fade. The earth is not an illusion a few paltry millennia in age; it has existed and changed for billions of years,

and from now on it is our responsibility. Surely this is a richer and more satisfying view of the world than that of the pure aesthete.

We could also turn the question around and look for another approach to analysis. Instead of seeking knowledge about the outer world, we could ask ourselves *why* we feel that something is aesthetically appealing. Why, for instance, is the countryside beautiful?

The attack on the search for knowledge comes from another quarter too. Our science and technology, it is said, have led to our poisoning and destroying our environment, threatening to annihilate it and ourselves as well. True. But in fact this has happened because our behavior is only pseudoscientific. We are turning a blind eye to elementary relationships, for instance in our predatory exploitation of natural resources. It is irrational to believe that in the long run we can take out more than we give back. No matter how sophisticated our technology, the thing is not feasible—at any rate not if we wish to keep the earth habitable for future generations, as most of us probably do. And it follows that we are not suffering from a surfeit of rationality but from a lack of it.

I recently got a letter from a reader who asked me if humanity is really "finished," whether we are not in reality stumbling about at the very beginning of humanization, and whether that is the reason we tend to behave so foolishly and seem unable to learn from our mistakes, no matter how often repeated. It made me think of a book by the Nobel laureate François Jacob, *The Possible and the Actual*—what could and what does exist. Possibly, in the

future, we shall find that the physical laws of the universe are such that the existing world is the only possible world, that a differently constituted universe simply wouldn't work, that Dr. Pangloss was right in saying that this is the best of all worlds because it is the only one possible. (Would that imply a Designer, a world Creator? The journalist and chemist Gustaf Mattsson tells a joke about the so-called anthropic principle, i.e., that everything is created for the benefit of Man—because, you see, if the hard granite were instead like semolina gruel, Man would not be able to walk on it.) According to the Talmud, there were twenty-six failed attempts to create a world, and only the twenty-seventh was successful. . . .

Even if this were true of physical law, it most certainly is not true of life. For all the forms of life that we see around us are but a small sample of the species that have existed during aeons of time—to say nothing of all the forms of life that *could* have existed. "The possible" in this case is an enormous spectrum and "the actual" just a few stock examples.

Has not evolution, then, during these millions of years, advanced toward ever higher forms of life? A look at the history of life gives a somewhat different picture. In fact, in the past evolution has created countless beautiful and seemingly perfected species that nonetheless died out long ago.

Nor can it be said that evolution is always "progressive" —at least not from every point of view. It turns out that evolution has no one direction but myriads of different directions; the main theme is not growth but diversification. There are instances that look like degeneration from

our point of view, such as the rise of parasitic organisms that may have lost most of their organs and come to consist essentially of a mouth, a gut, and sexual organs. Here we may establish as a fact that evolution does not point "forward." It does not point anywhere: it is utterly opportunistic. No optimistic view of the future can be based on Darwin, as he himself well knew. Biological evolution does not go upward; it just goes somewhere, it is amoral. Assuming a congruence between biological and cultural evolution is ridiculous. The very mechanism differs in the two cases: natural selection (Darwinian) in the former, "inheritance" of acquired characteristics (Lamarckian) in the latter—that is, a cumulative growth of experience. This may sound complicated, but why should we imagine that the world is easy to understand?

How do you create a perfect organism? An engineer who wants to create a machine will make careful plans and use the best materials that can be obtained with the help of modern technology, and will be able to calculate beforehand what the machine can do, how much it will cost to build, and so on. This parallel was in fact drawn by pre-Darwinian biologists, the most famous one being the divine William Paley, author of *Natural Theology* (1802). It should be noted here that Paley was by no means a reactionary, but, on the contrary, expressed himself sarcastically about "the divine right of Kings," which he compared with "the divine right of constables." That made him largely unpopular within the Establishment, which was dismayed by his "freedom of expression." It is too easy to divide people into bad guys (those who were wrong) and good guys (those who were right). In many cases it is simply a question of scholars who were well versed in the science of their time and who did excellent work, sometimes against the current of contemporary

thought. To return to Paley: his argument was that the creations of nature are so perfectly adapted to their mode of life that this must prove the existence of a plan, i.e., direct creation. Joshua Lederberg calls this theory "instructivism."

Our organs, however, are *not* perfect. Our eyes, to take an example, are really absurdly constructed in that the light-sensitive cells of the retina are turned away from the light, an oddity we share with all other vertebrates. The octopus has eyes, independently evolved, that are very much like ours except that the light-sensitive receptors are turned the "right" way, which is a more efficient arrangement. Of course, we have other weaknesses, for instance the fact that our upright position often results in back pain.

That is because we did not get where we are through instant creation but through evolution. And evolution behaves very differently from the engineer. François Jacob compares it with "tinkering" (*bricolage* in French). While the engineer works with the best available materials and with tools especially designed for the work, the driving force of evolution, natural selection, works by the tinkerer's method: the tinkerer uses whatever is at hand—old cardboard boxes, string, bicycle parts, God knows what—to put together something that works, although not always as expected. And, of course, natural selection has

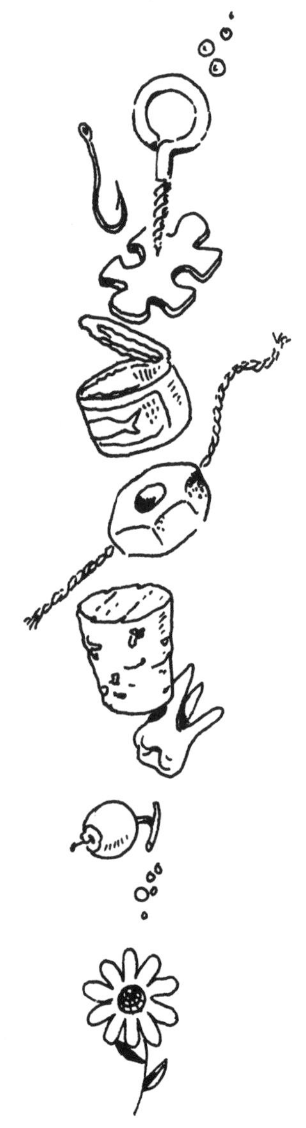

to work with what is at hand. (An aside: to exterminate a species of animal or plant implies that we kill something that is "at hand," in our hands, and thus we impoverish not only the present but also the future.)

This "tinkering" Lederberg calls "selectionism," which is the opposite of "instructivism." Evolution works like a tinkerer who, during millions of years, slowly modifies products, adding or subtracting something, changing, turning this way and that, always according to the demands of the current situation. Evolution never acts with a view to the future; it functions entirely opportunistically in the present.

Snake eyes give us a good example of Jacob's *bricolage*. Snakes evolved from quadrupedal lizards during the Cretaceous period, some 100 million years ago. Their eyes, however, differ from those of normal lizards by poor development of the lens and retina and the absence of eyelids. This suggests that the first snakes evolved from digging, subterranean lizards that had become partly blind.

To restore functioning eyes in the snakes, which returned to life aboveground, evolution had to tinker with what was at hand, and did produce acceptable organs of sight. Still, what had been lost could not be completely regained.

Now it is easy to understand why humanity is not "finished." We are not the result of a plan. We have originated through natural selection's tinkering with the material at hand, and we ourselves constitute material for what may come in a future that we, perhaps, in spite of everything, may dare to hope for.

Does selection still tinker with us? Yes, certainly. Let us take an instance: our love affair with alcohol. It has a tradition that goes back many generations, and has without doubt entailed an ongoing selection for increased resistance. Still, there are many who suffer damage from alcohol, and this means that selection is still working—for instance, in making the liver more efficient. (If everybody drinks, those with poor liver function die young and have few offspring, while those with cast-iron livers get on well and have many children.) That this can happen is proved by many bird species, for instance the waxwing, which in the autumn may eat fermented rowan berries and as a result become dead drunk. However, it has a

really efficient liver and recovers quickly, presumably with minimal hangover. In a similar manner, we may expect that a general use of cannabis would usher in a selective situation that started with much suffering but in the end improved our resistance. I doubt it would be worth it.

The way in which we experience the world is also a result of selection's tinkering. We have our "five senses," which in reality are at least eight (taste, smell, sight, hearing, tactile sense, a sense for pain, a temperature sense, and a "position sense" which, for instance, lets you know without looking that your arm is bent). Many vertebrates also have an electromagnetic sense. Through our eight senses we experience the world. That we have exactly these senses has a historical explanation. At the fish stage we evolved sight, smell, taste, and hearing (the tactile sense is surely still older). Moreover, we then had a unique "fish" sense, localized to the lateral line system, that seems to have to do with the sensing of very long-wave pressure oscillations. It happened that these senses were sufficient for the fish to get along in their world (as far as we know—plus the pain, temperature, and position senses). The lateral line sense is not present in land vertebrates except for early amphibians—presumably it does not work in air—but the others seem sufficient for us to get along. However, the ambient world we experience is determined by these senses and their receptivity. We have no sense registering radioactivity, and so it exists to us only as an abstract concept: we can observe its effects but we cannot "feel" it.

On the basis of sensory information our brain forms its own picture of the ambient world, a picture that is simplified and biologically useful but very far from an exact replica. Many animals experience their world in a quite different way: the dog as a world of smells, the bat as a

world of sonar echoes, the bees as a world without reds but with ultraviolet colors, and so on. We are all prisoners of our senses. And yet, in an evolutionary perspective, the world of our senses is not static: the senses themselves may evolve, and so does the brain that receives the information. Our image of the world also results from a long evolution, in which the things that are important for survival have been retained, while those things that were momentarily unimportant may have been lost.

We may note, for instance, a loss that has developed into tragedy for many people. We are unable to synthesize vitamin C in our bodies, and lack of it causes us to die of scurvy. This can be explained from the fact that our ancestors among the apes were fruit eaters and thus got such large doses of vitamin C that any addition was unnecessary, or even harmful (we rarely think of vitamin C in terms of overdoses). Exactly the same kind of selection has evidently operated in swine, which, similarly, are unable to synthesize vitamin C.

How about what used to be called the "sixth sense," or telepathy: thought reading (also over long distances)? Telepathy has many staunch believers, who think that at least some people have the ability. On the other hand, it has never been possible to demonstrate its existence with acceptable scientific documentation (as has been clearly shown, for instance by the dramatic unmasking of various impostors by James Randi of the CSICOP). And there is really no reason to believe that telepathy does occur, on the very simple rationale that the faculty would be of such enormous value that natural selection would rapidly have made it as general and self-evident as the faculty of sight. That communication between human beings has enormous survival value is evident from the fact that our speech apparatus is a miracle of efficiency and complexity—unlike its corresponding organ in other mammals such as the apes. We have developed the speech organs so far that we are at risk of choking on our food and drink! Selection, in other words, has set down a gigantic work in finishing this exquisite instrument. How much more precious would not direct thought reading be, were it possible? The least rudiment of it would have been seized upon by selection long ago, if it had occurred in human beings, or in any other animal for that matter.

One evolutionary adaptation is that we have learned to *like* our environment—just as we think orange juice tastes good. For millions of years, life must have been unbearable to the few oddballs who did not like the country. We are predisposed to like the country. And that is the answer to the question heading this chapter.

In a work of art that at first glance seems chaotic, closer scrutiny may reveal lines or color patterns that give us joy. Chaos is "ugly," but when some kind of order emerges it becomes "beautiful." This is trivial enough, yet it also points to something lurking that is good for survival. We like succeeding; we hate to be failures. In a chaotic world we are failures, helpless; we have a basic need for orientation in our existence, and without that we could not survive. The play of our emotions is adaptive and has evolved by selection. We "like" to behave in a way that ensures the survival of the species.

That we are also at home in an urban environment, although it may be poles apart from the countryside, may in part be due to the fact that the town's pattern of streets

and blocks is fairly logical and easy to survey. Even without a map we quickly form a mental chart during a walk in an unknown city. The landmarks in fact are overly distinct compared to those in nature—the buildings, squares, street scenes have great individuality, at least in older townships. The discomfort felt in the ultramodern city may perhaps be partly due to the monotony of the architecture.

Of course, that is not the whole story.

We humans are extremely social animals, which is also the outcome of a long evolution. To a being like us, a positive relationship with our own species is of great importance to survival, and it is evident that this is deeply programmed into our psyche. Humanity is humanity's joy.

In this case it means that we want to see people in the town. A ghost city seems scary because there are no living humans. On the other hand, because we like to shudder, impresarios sell ghost towns to tourists with considerable success. Nothing in biology, especially human biology, is simple.

One facet of our social need was found by Darwin to be so important that he made it the main theme in his book on the descent of man. That was the relationship between the sexes. Darwin thus discovered *sexual selection,* which means that everybody strives to find a sexual partner who is most attractive to that particular individual. (This is not, it may be averred, a new discovery as such,

but it was unusual in a scientific context!) It is interesting to reflect on the meaning of sexual selection. For hundreds, thousands, millions of years men and women, and before them male and female hominids, have been at work selecting a kind of ideal of the opposite sex. When men or women speak denigratingly of the opposite sex, therefore, it is really themselves, their own sex, that they are rejecting; for it could be said that woman was created by man, and man by woman. Perhaps the situation we are in is just the one we deserve. . . .

CHAPTER FIVE

Thinking Biologically

To most people biology is a fairly innocuous subject having to do with hobbies, such as nature walks or bird-watching—when it doesn't evoke the menacing specter of genetic manipulation or wistful thoughts of a less polluted world. Physics and astronomy seem to dominate the "real" cultural debate, to the extent that the sciences are considered at all (deplorably, not much), although, as the physicist P. W. Anderson commented perceptively in 1972, "The more the elementary-particle physicists tell us about basic physical laws, the less relevance they seem to have for the exceedingly real problems of the other sciences, not to mention those of society." Perhaps he exaggerates, but there is no doubt that the debate is one-sided, and biologists themselves must remedy the situation.

In 1982, the famous ornithologist and authority on evolution Ernst Mayr published his magnum opus *The Growth of Biological Thought*. That book is crucial for our approach to and our understanding of the history and philosophy of science—not least our understanding of the world view offered by modern biology. Unhappily, it has hardly been noticed in the debate on the subject. I suspect that this is due, at a minimum, to the fact that the philosophers of science—those who belong to the profession as well as the amateurs—generally begin their studies in physics and the other so-called exact sciences but lack any biological training. (One example is Popper, who because of his minimal knowledge of biology eventually made serious errors, which he himself admitted.)

In the opening chapters I have given examples of various biological approaches to literature and art. But what does it actually mean to think biologically? Not least, it

gives us the ability to banish various kinds of misunderstanding, as exemplified by the often-heard disparaging term "biologism"—not to mention the totally unacceptable "social Darwinism" for what ought rightly to be called "Spencerism." It was Herbert Spencer who applied Darwin's theories to the social sciences, where of course they do not belong, a fact of which Darwin himself was fully aware. His disapproval of Spencer's ideas is clearly shown in his letters, and it seems unethical to associate Darwin's name with a theory from which he disassociated himself.

Mayr is one of the leading figures of modern biology. Together with the zoologists Bernhard Rensch and Julian Huxley, the geneticist Theodosius Dobzhansky, the botanist Ledyard Stebbins, the paleontologist George G. Simpson, and some others, he is one of the chief architects of the modern synthesis of the theory of evolution, and thus is especially qualified to speak for the science of biology. Although biologists of different backgrounds have sharply diverging views, which is only to be expected in a dynamic science, a core of concepts and principles does exist that is adhered to by the great majority of biologists working today. They are the ones who concern us here.

Essentialism and Population Theory

Essentialism, a school of thought that dominated scientific theory for more than two thousand years, goes right back to Plato. To him the existing world was merely an imperfect reflection of various ideal forms, which he called ideas (*eidos,* a word having the same root as our "idea") and which the Thomistic philosophers gave the name "essences." Biologically, this implied, for example, that every species is an embodiment of the immutable essence of the species: individuals are not exactly like one another because of imperfections in the way they are reproduced. So deeply rooted was this kind of thinking that not until Darwin did thinking based on populations as a whole gain a foothold.

According to the population view, everything in the organic world is unique; the most important thing is the individual, not the "type." The "average person" is an abstraction from the real world, not an idea imperfectly reproduced in the form of real "average people." Individuality and uniqueness exist on all levels in the organic world, ranging from genes, cells, and organisms and their interrelationships to the whole of life. Each bacterium is unique; no other is exactly like it. Identical twins do have identical genes, but they are still unlike each other—unique—since the influence of the environment can never be the same at all times (not even in the womb). The same holds true, to an even greater extent, for the more complicated

biological differences. That is why entities in biology must be studied in quite a different way from groups of identical inorganic entities, such as, for example, those considered in terms of atoms or molecules. Such is the basic thinking behind the population theory. The differences between individuals, for instance individuals of the exact same species, are real and significant, while in the view of the essentialist they are the result of "errors." Variability —the extent to which it is measured—is thus one of the most important parameters in population theory, while to the essentialist the word is a meaningless noise.

Essentialism in biology leads to *typology,* meaning, for example, that every species—or race—appears as a distinct, unchangeable "type."

Although essentialism in biology has been replaced with population theory on the whole, a kind of vulgar or low-brow essentialism survives, and exerts a most unfortunate influence on ethics. Most of the assertions made in racist literature reflect a kind of essentialistic thinking (typology: a negro is a negro is a negro). In the same way, essentialism is a sexist doctrine and serves as the basis for such views as that women should stay close to the hearth and bear children. Of course, this has nothing to do with modern biology. Most such totalitarian views belong, obviously, in the dustbin of history.

The introduction of population thinking was Darwin's greatest contribution to philosophy. It led to a completely new way of observing reality. In turn, it led him to the discovery of the law of natural selection.

Reductionism and Emergentism

Biology as a separate science has been unpopular among certain physicists and physics-oriented philosophers of science. The distinguished atomic physicist Ernest Rutherford compared biology to "stamp collecting." Such arrogance shown toward unfamiliar sciences is not unusual, and represents, as I pointed out earlier, a biological phenomenon, none other than attempts to preserve one's territorial rights. (I am not saying this to disparage Rutherford's immense contributions to science. In fact, a paleontologist has special reason to be grateful to him, since his work in large degree contributed to the dating of the earth's age.) Therefore, it has been maintained that all theories of biology can, at least in principle, be "reduced" to theories of physics, which would restore the unity of science.

But here we must stop and analyze our concepts, since "reduce" is far from an unambiguous term. Mayr differentiates between three types of reductionism.

The first, which he calls "constitutive reductionism," asserts that all organisms are composed of the same kind of material substance that we find in the inorganic world, and that organic processes are fully compatible with the physical and chemical phenomena that one finds at the level of atoms and molecules. Reductionism in this sense may be said to fall completely within the scope of biology. The contrary position is taken by so-called vitalism, according to which there are certain processes in living organisms that do not obey the laws of chemistry and physics. But vitalism, like essentialism, is a dead horse in modern biology.

Something quite different is the "explanatory reductionism" according to which one cannot understand a whole unless one has dissected it into its various components, and these in turn into theirs, until one arrives at the ultimate, basic structures—in biology, to the molecules. This process in many cases is important, as it was, for example, in learning the functioning of the genes. But in other cases the method is as unreasonable as describing a Beethoven symphony as a sequence of carefully registered sound frequencies. Extreme analytical reductionism will usually be meaningless, since it cannot give any idea of the interplay within a complex system; an isolated organ, for example, has distinctive features different from those of organs in a functioning organism. I shall return to this subject in my discussion of emergentism.

The third type is "theory reductionism," which maintains that the theories formulated within a field of science (with complex entities), can be seen as special cases of theories and laws within another field (one whose entities are less complex). Thus, you should be able to reduce biology to physics by deducing the laws of biology from those of physics. In fact, that is neither possible nor desirable. Biological processes are physical and chemical processes, but phenomena like natural selection, species, cell division, territory, and courting are biological *concepts,* and a purely physical description of them is irrelevant. So this type of reductionism does not belong in biology.

In most systems, at least the organic ones, the characteristics of the whole cannot be deduced, even in theory, from the most complete knowledge of its various components. The appearance of new characteristics in a whole is called *emergence.* Emergentism is, philosophically, materialistic and must not be confused with vitalism. What emergentism claims, quite simply, is that "explanatory reduction" does not give us the whole answer and that complex systems have to be studied on all their levels, since each level possesses characteristics that are not apparent at the next lower level. We cannot explain the makeup and functioning of a human body merely by studying its cells, any more than we can describe and explain a society merely by studying its separate individuals. In all such cases, the interaction between the components will lead to new characteristics manifested by the whole.

For scientists, this situation makes it necessary to decide for ourselves on which level we will proceed with research, whether in molecular biology, genetics, physiology, anatomy, ecology, animal geography, etc., in order to illuminate the system that is being studied. That also implies a decision to leave certain black boxes unopened. Thus Darwin himself left the question of the causes of variation unsolved when he published his theory of natural selection.

Experimentation and Observation

Scientists who devote themselves mostly to experimentation claim that experimentation is the only proper scientific method. Experiments are also of central importance in biology, especially within the field called functional biology; but within biology as a whole we find that the comparison of observations is even more important (just as in astronomy, geology, meteorology, not to mention the humanities). Actually, the difference between them is not that great. Experimental scientists observe their own experiments, while the others observe the experiments of nature. Both are "descriptive" scientists, although many experimentalists would like to limit this (disparaging) term to those who deal in comparative observations.

Actually, both methods are essential. To mention a few examples: ecology, paleontology, and behavioral research (etiology) are based to a great degree on comparative observations but also on experimentation. (That is true even of paleontology!) To make meaningful observations, however, one must be able to classify the objects of one's

research; this was realized as early as the early 1800s by Georges Cuvier, the pioneer in this type of research. It is no accident that Harvard's renowned research institute, to which Mayr himself is connected, is named the "Museum of Comparative Zoology."

The Theory of Evolution

The theory of evolution is an example of a doctrine that was created almost entirely on the basis of comparative observations. Darwin did do experiments, true, but they played a minor role in the fashioning of his whole theory.

Evolution is progressive (leads to advances); or, Evolution is due to chance. These are the two most frequently heard—and mutually contradictory—misconceptions about the biological theory of evolution. They seem to be ineradicable, not least because of the inability of the leading philosophers of science to find out enough about biological issues, a situation that has only begun to be corrected in recent years through the work of such philosophers as Ruse and Sober. In the following, and mainly on the basis of Mayr's *The Growth of Biological Thought,* I shall attempt to set out the most important principles underlying the modern theory of evolution.

The diversity of life is enormous; it is estimated that at the present time alone there are millions of kinds of organisms. The number of species that have existed during the long history of the earth has been estimated at several billion. To study this enormous diversity at all, we need a system of classifying its components. The science that concerns itself with this is called systematics. To Aristotle, such classification mirrored the "harmony of nature"; to natural scientists like Linnaeus, it unveiled the "plan of the Creator." These metaphysical interpretations were replaced with a scientific interpretation through Darwin's theory of a common origin of species. A modern biological taxonomy attempts to construct a hierarchy of units in a system (i.e., species, genus, family, etc.) that correspond to the degree of real, genetic relationships. Although systematics of this kind is the oldest branch of biology it is still changing rapidly.

Darwin's theory of evolution is in fact a complex of five different theories that have met varying fates but are all fully accepted by modern biological science.

The first theory maintains that the world is not immutable but is subject to continual changes. Of course, this is not easy to accept for an essentialist but pretty self-evident to adherents of population theory. That the world is changing and changeable, and on a grand scale at that, may now be considered an established fact rather than a theory. It is proven, for example, by the changes one can observe in geological layers—areas that are now dry land in the past were completely covered by the sea, and vice versa; species of animals and plants have become extinct and have been replaced by other species; and so on.

The second theory claims that all organisms living today have a common origin and have developed from common primitive forms through continued division into different species. By including human beings in this "tree of evolution" Darwin deprived them of the privileged position that until then they had enjoyed in nature's scheme of things. The loathing of this conclusion caused an uproar that has still not abated. But the theory of a common origin was quickly accepted by all leading biologists and is now regarded as self-evident. (This is not the place to deal with the overwhelming evidence proving the correctness of the theory.)

The third theory has it that evolution occurs gradually, not in sudden jumps, and it quickly faced much opposition. The theories of evolution by leaps and bounds (saltation theories) were put forward even in Darwin's day and had great influence at the turn of the century. By now

they have been largely discarded (except for allopolyploidia, the process through which the genes of two different plant species lead to the creation of a new species when they are combined). At the present time there is a certain theory according to which new species will often appear "suddenly"—that is, in a geological perspective—but even that theory maintains that this occurs through an at times radical increase in the speed of evolution, not through any "leaps." Even Darwin pointed out that the tempo of evolution may vary.

In his fourth theory Darwin dealt with speciation, that is, emergence and the origin of two (or more) subspecies from one extant species. Since today's enormous diversity among the world's organisms has come about through speciation, that is surely one of the most important evolutionary processes. Darwin explained it with regard to populations (groups) by pointing out that two populations of one species will slowly develop in different directions; and this theory has also proven to accord with observed facts.

Darwin's fifth theory, which biologists now refer to by the term "Darwinism," is the theory of natural selection, the cause of evolutionary changes. (It was launched by Darwin and Wallace unknown to each other. Darwin's version was broader and more thorough as well as more convincing.) Although this theory is central within modern biology, almost like the atom theory in chemistry, it has remained the most misunderstood of all the five building blocks that make up Darwin's complex of theories. It was a theory that Darwin's contemporaries found hard to swallow. Not until the beginning of the modern theoretical synthesis of evolution as a whole has it become fully accepted. It is now considered Darwin's most important discovery. But only very lately have the philosophers of science come to realize what it implies.

The most common error is to view natural selection as inherently progressive. Science and technology are progressive (they make advances), and perhaps by analogy with them, it seemed natural to regard evolution too as progressive. But science and technology are goal-oriented activities, in contrast to evolution. Of course, we can define "progressive" in different ways—as, for example, "more like humans" (more intelligent, with greater control over the environment, etc.), and when we then consider the evolutionary line that has led to us humans, it does seem to be progressive. But it is merely one evolutionary line among the billions that have existed during the 3,500 million years of our long history. According to the same criteria, many other evolutionary lines appear to be regressive, while still others are indefinable. (Which one is more like humans, the crayfish or the crab? The latter has evolved from ancestors similar to crayfish.) No matter how shrewdly we define the word "progressive" we will be caught in the same bind.

It is not possible to provide an exhaustive description of natural selection here. Briefly, it revolves around the idea that certain individuals in a variable population, because of certain traits, will have a larger number of offspring than others, and so these traits will come to predominate in the population to an ever greater degree. The manner in which the selection is effected is not strictly deterministic but rather probabilistic.

Darwin did not know the causes of the variation within populations, although this variation is clearly necessary for selection to take place. In his version of the theory it remained a black box. We now know that changes in hereditary factors come about through mutation. We also know that mutations are "accidental," in the sense that they are not related to the actual needs of the organism. On the other hand, they are not accidental if we mean by this that they can lead to any result at all, such as changing a rabbit into a bird.

We might easily be tempted to conclude from this that mutations account for the rise of variations. But in that case we would be overlooking the fact that in populations with sexual reproduction, it is the *recombination* of the genetic material above all that leads to variation. An oft-cited arithmetical example may help explain this argument. Suppose that in a population with asexual reproduction (as, for instance, bacteria) ten mutations occur. The result, then, will be ten new genotypes. But in a population with sexual reproduction, where what is inherited from both parents is combined, the potential result is three raised to the tenth power, or close to sixty thousand new genotypes. The base number is three because of the fact that two versions of the same branch—the original (A) and the mutated (a)—may be combined in three different ways (AA, Aa, and aa).

All this grows more interesting if we cast a glance at the earliest forms of life on earth. The first living things made their appearance about 3,500 million years ago. They were kinds of bacteria (part genuine bacteria, part blue-green algae), and they reproduced asexually. As a result, very little happened for billions of years. The first evidence of sexual reproduction (in connection with more advanced one-celled animals) is known from the period 1,500–1,000 million years before the present. After that the speed of evolution shot up.

It takes 100 mutations (in a species with sexual repro-

duction) to produce as many possible genotypes as the number beginning with 51 . . . and followed by 48 zeroes. Since normal populations represent thousands of mutations, it is not hard to see that the potential for diversification reaches astronomical figures. According to one calculation, the total number of genotypes within the human population—and thus the total number of conceivable unique individuals—exceeds the number of atoms in the known universe. A small number of mutations thus suffices to give the selection process an enormous amount of material to work with.

The selection process functions opportunistically and normally in the direction of increased adaptation to actual conditions. (Nonadaptive selection is limited to certain types of sexual selection.) In changed conditions the selection process may change its direction. Selection is the orientating factor in evolution. To call evolution "accidental," as do the Creationists and other pseudoscientists, is therefore so much nonsense. One might compare the variation within a population with the potter's clay and the selection process with the hand that throws and shapes the material.

The selection process functions even on the inorganic level—for example, when there is a question of growth (such as the growth of crystals) and when there is disparity among those who survive (for example, certain kinds are more durable than others). On the organic level it is completely dominant.

The zoologist Richard Dawkins gives a humorous example, based on the possibility that a battalion of monkeys punching away at typewriters might in time succeed by pure chance in writing all Shakespeare's sonnets. Dawkins limits his hypothetical experiment to the writing of a single sentence from *Hamlet:* "Methinks it is like a

weasel" from act 3, scene 2, where Hamlet is talking to Polonius about a cloud. The sentence requires 28 keystrokes (including the letters and the intervening spaces); the English alphabet has 26 letters. Including the keystrokes for spaces between words, the probability is that one monkey will write down the words correctly in a number of attempts represented by 1 over 27 to the 28th power (a number that has to be written with 40 zeroes). Even with the use of a computer, which can spew out complicated solutions in seconds, the entire age of the universe would not give it enough time to arrive at a correct answer to this problem. If, on the other hand, the computer is programmed to start with any nonsensical sentence at all, such as WDLMNLTDTJBKWIRZ-RELMQCOP, and then produce a number of accidental "mutations" per generation and thereafter always select the "child" that most resembles the intended meaning, it will arrive at the correct sentence, METHINKS IT IS LIKE A WEASEL, in slightly more than 40 generations. The example shows the power and importance of selection, but in this case it is not at all biological, since it attempts to reach a preconceived outcome.

The example of the potter's clay is illuminating in that it brings home the *creative* and *originating* role of selection. This seems to be difficult for nonbiologists to grasp—including professional philosophers. At the very most they will accept that selection has a purging function—sifting out "inferior" genes. They overlook the most important function, which is to build up favorable *gene combinations.* Selection is creative to a much greater extent than it is purgative. Natural selection is the only process that can do this, and that is why it should be regarded as the creative principle of evolution. There is none other—nor is there any need for one. The law of natural selection must be seen as a law of nature, which follows logically from certain known facts but which could have been discovered only on the basic rationale of population thinking. It is valid for all living things, and also for any possibly existing extraterrestrial life.

Is selection effective? Observation as well as theoretical research answers in the affirmative. A combination of genes that results in an advantage in the selection process of 1/1,000 (and thus results in progeny of 1,001 individuals instead of 1,000), within a relatively small number of generations will completely dominate the population, even though its numerical advantage may seem hardly noticeable. In regard to its material, the selection process functions perfectly.

Selection always operates solely on the *level of individuals* (according to Mayr and most of the other researchers in the field of evolution), not on the level of genes or groups, in spite of such sensational theories as the one based on the "selfish gene" and another one positing "group selection."

An interesting circumstance, which has not been demonstrated until quite recently, is that the genetic material to a large extent is "dead," that is, it has no effect on the nature of the individual. Thus, mutations in such genes are "neutral" and are not influenced by the selection process. Since they do not have any influence on an individual's traits, they are of no practical importance to evolution as such. The neutral mutations, however, are of interest to those who carry out research in evolution, since they become concentrated in the course of time and hence constitute a kind of molecular "clock" that provides a measure of the tempo of evolution. There is a possibility that such "dead" sectors may be activated by so-called regulator genes, and thus may be seen as being held in reserve.

Selection produces "teleological" effects, which implies that it appears to be goal-oriented (teleology is the doctrine about a predestined goal). Even Plato thought that one can discover a purpose in nature and in natural events. The nature theologians felt that the same scheme could be detected in the harmony of nature and the innumerable adaptations displayed by the various organisms. They were regarded as a conscious plan of creation. After the acceptance of the theory of evolution they switched over to seeing a conscious plan in a "progressive" evolution.

As Mayr demonstrated in 1974, the term "teleology," unlike the familiar insect, can be dissected into not three but four different parts.

The first one is called "teleonomy" and is exemplified by a genetic program that governs our individual development from the egg to an adult person. The term also involves physiological and behavioral processes that are governed by an inborn program. (The idea of such programs is a new one but became generally and seriously accepted at the time the computer was invented; this is an example of a debt of gratitude that biology owes other fields of research.) All the biological programs, of course, have been created and polished by evolution.

From this we speak of the teleomatic processes, a term that covers the fact that a definite endpoint can be reached in accordance with physical laws. A falling stone that hits the ground and stays there is a good example. In a biological context this is ordinary and trivial.

The third concept consists of *adapted systems.* The eye has been constructed for the purpose of seeing, the brain for the purpose of thinking, the kidneys for the removal of waste products. Darwin showed that these systems had developed through natural selection. The term "teleological" is therefore misleading.

Fourthly, we have "cosmic teleology," that is, the theory that the entire universe and all its components have developed in accordance with a cosmic plan. This type of teleology is metaphysical and so has nothing to do with science and does not concern us here.

Natural selection tends to lead to adaptation through changes in the genetic programming. (In certain cases, the importance of which, however, is disputed, selection's functioning may be disrupted, when populations are so small that accidental changes overshadow selection.) Increased adaptation, however, cannot be equated with "progress" in any reasonable sense. The idea of a *scala naturae* (nature's ladder) from the lowest to the highest is simply irrelevant to living nature. The distinctive stamp is not progress but rather the origin of diversity—uncounted myriads of different forms of life, each one adapted to its unique niche in life over billions of years.

Thus evolution has four great themes.

Continuity. Many species remain unchanged through millions of years—presumably when they are well adapted to their environment and natural selection functions so as to preserve them. Innumerable examples of this phenomenon, which is called *stasis,* are known in the science of paleontology.

Change. Species change in the course of time, and in the end we will have a population in a future generation which is so unlike the original form that it must be characterized as a new species. Innumerable examples of this are known from the fossil record.

Multiplication. A population can be divided into two or more populations, each evolving in a different direction and giving rise to several species. We know this course of events best in some detail from our now living organisms, which show all the stages in the branching-out process; in broad outline this process is always present in the fossil record.

Extinction. Of all the various species that have lived on this earth the overwhelming majority have become extinct, without any progeny—and this is completely indepen-

dent of whether they have been the most advanced or primitive ones, from our viewpoint. The species now existing certainly make up fewer than one in a thousand of all those that have ever existed.

But no matter which theme was valid in any one time or place, it is selection that is dominant, selection, which creates the new and destroys the old. It is the omnipotence of selection that we encounter in the world of life and that we see all around us, and within ourselves.

Evolution is like poetry: it explodes into forms and colors that are always new, without any overlapping goals or meaning, always individualistic, always shaping and reshaping; it creates, it plagiarizes, it annihilates.

CHAPTER SIX

The Clock in the Rock

When geologists talk about the age of some rock or other, the thoughtful listener is apt to become somewhat glazed of eye, rather like when the media report on, say, the U.S. military budget. The numbers are so immense that it is hard for us to take in what they actually mean. We can *say* that the earth is upwards of 5 billion years old (5,000 million), but can we actually *visualize* what this means?

To illustrate the history of the world it is common practice to scale it down to twenty-four hours and then describe what happened in the morning, in the afternoon, and during the last seconds (when humanity appeared). But this is just a desktop model that can give no feeling for the cosmic scale of the events.

Alternatively, you could use parables purporting to give a concrete idea of what a billion is, or for that matter a million. By the way, if you want to see a million, probably the best way is to get a roll of graph paper divided

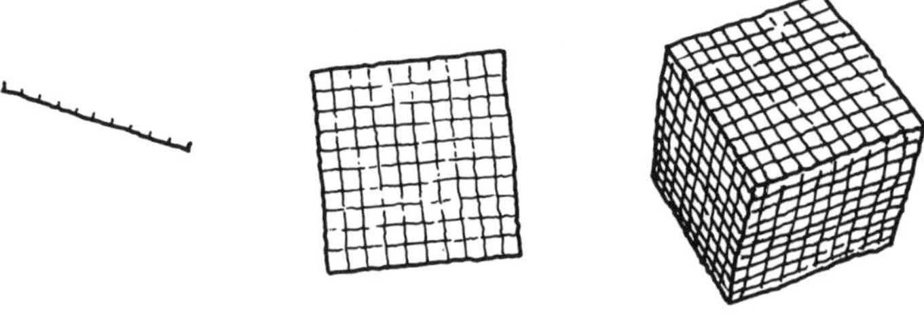

into square millimeters and cut out a piece one meter long and one meter wide. It will contain precisely one million small squares. If you lay out one thousand such sheets end to end—which makes one kilometer—you can let one billion small squares pass before your eyes during a ten-minute walk.

Or let us make a birthday cake for the earth, a cake with five billion candles (if you feel niggardly, 4.6 or 4.7 billion may be a better estimate). If we place the candles at a distance of two centimeters from each other (a little less than one inch), we shall need a cake covering some two square kilometers (200 hectares, or 494 acres) to accommodate them all. Many farms, at least in Europe, are a lot smaller than that.

Or imagine an hourglass that lets through just one grain of sand in a year. The total will still amount to a decent truckload of sand.

Another possibility, which was once suggested by Milton Hildebrand, is to compare geological history with cosmic events and dimensions, such as the rotation of galaxies and the emptiness of space between them. This may be coupled with the feelings of awe that we experience when, on a starlit night, we gaze into the incredible distance of faraway worlds.

700m
... 749.892.978- ...
749.892.979- ...
749.892.980...
... 749.89
750m

You can play with descriptions in this manner but perhaps will not get very far after all. We simply are unable to grasp fully such stretches of time—they are beyond the reality in, and for which, we evolved.

How do you measure this abyss of time?

This question is usually answered with a description of the radioactive elements and their gradual transformation into other elements. For instance, a radioactive isotope of potassium (which is a very common element in rocks) changes into either calcium (in 89 percent of cases) or into the gas argon (in the remaining 11 percent) at a constant rate, so that of a given amount of radioactive potassium only one-half remains after 1.3 billion years, while the other half has changed into calcium and argon. After another 1.3 billion years, only one-half of the one-half of the original amount of radioactive potassium remains (that is to say, one-fourth), and so it goes on to one-eighth, one-sixteenth, and so forth. By measuring the amounts of potassium and argon in the rock, we get a measure of its age. (The calcium would in theory do just as well, but unfortunately it cannot be differentiated from other calcium already present in the rock, so it is useless from this point of view.) This method is called *radiometric dating*. It does not have to be potassium—there are other radioactive elements that can be used in a parallel way—but the principle is the same. Radiometry has given us a firmly based time scale for the history of the earth.

Well and good. But the lay reader may object that this is an esoteric method and that it would be rather nice to see something simpler—more obvious. For instance? I reply. Something I can see, the reader might say; something like tree rings.

Yes, why not? The earth does have annual rings, just like trees. And the interesting thing is that they give time

scales that are in excellent agreement with the radiometric ones.

The first to be developed was the glacial clay "varve" or band chronology, which was introduced by the Swedish geologist Gerard De Geer. It makes use of the fact that glacial clays (which form in meltwater lakes by an ice margin) show a rhythmic alternation between light and dark sediments, which are seen as bands in a cross-section. The light bands are formed during the summer, when melting is brisk and the streams bring along large amounts of sand and clay particles. The dark ones are laid down in

12.797!

the winter, when the streams cease to flow and only fine detritus is precipitated to the bottom.

If the summer is warm, melting is prodigious and a lot of material is deposited, so that the light band becomes thick. In a cold summer, conversely, a thin varve is formed. So there appear characteristic sequences of thicker and thinner varves, which you can identify from place to place. It was indeed necessary to go from place to place, for the ice margin retreated to the north and northwest as the Pleistocene ice-sheet melted off. And so the sequence of thinner and thicker varves that was at the top of one varve series could be matched by a corresponding sequence at the base of another series farther north. By combining such local sequences it was possible to produce a chronology for the end of the Ice Age when the inland ice melted away.

At the beginning, it was a "floating" chronology—that is, it was not connected with the present day. This has now been rectified, especially by studies in the valley of Ångerman River in northern Sweden, where Ingemar Cato was recently able to close the last remaining gap. And so there is now an unbroken sequence of annual varves taking us nearly 13,000 years back in time.

The glacial varves, which may be a centimeter or more in thickness, are usually easy to observe. But there are also much thinner varves, perhaps only a fraction of a millimeter in thickness. Such varves, or lamellae, are formed in the bottom sediment of certain types of lake, which have been studied especially in Finland. This happens in "poor" lakes (technically termed oligotrophic) where there is no bottom fauna to poke about in the sediment. When studied microscopically, the varves are, again, seen to result from seasonal changes—the spring flowering of microscopic plants called diatoms, changes in the chemistry of the water, and changes in the supply of inorganic sediment.

They too are annual "rings" that can be counted and used in a chronology (if the lake is still in existence, the chronology of course will extend all the way to the present). The difference is that all the varves are piled on top of each other in a single lake basin, so that you do not have to move from place to place to construct your time scale, as is the case when you track series of ice-dammed lakes that followed the retreating ice margin. In this case, certain lakes too give us a long chronology; Lake Valkiajärvi in central Finland, which was studied by Matti Saarnisto, carries us back almost 9,500 years in time, to the birth of the lake after the melting away of the inland ice in this area.

It might reasonably be expected that such lakes existed in time past too—having vanished long ago but leaving a fossil lake bed perhaps protected by other, covering sediments. And so it is; the number in fact is legion, and we can take note of only a few examples. Let us start with a look at the lakes that existed during the Ice Age, or more precisely, during the interglacial periods—times when the climate was as warm as it is now. (We, of course, live in another such interglacial period, the warmest part of which is already long past—and which, as Ångerman River and Lake Valkiajärvi teach us, has lasted for about 10,000 years.)

Fossil lake bottoms of the same type—but overlain by "cold" sediments laid down during a glacial age—are known in various parts of the world. In Germany the annual varves of one of these lake bottoms were counted, while at the same time, by analysis of the fossil plant pollen in the sediment, it was shown that the entire history of the interglacial was recorded there. The record starts with a tundra vegetation, shows the immigration of

birch and pine, attains a culmination with oak and other broad-leaved hardwoods, and then returns by stages to the tundra. All this, as shown by the Quaternary geologist H. Müller, was enacted within 11,000 years—a length of time comparable with that of our current interglacial.

The chronology, of course, is again a floating one. All we can say on this basis is that it dates back some considerable time, because it was followed by a cold climatic oscillation, which in turn gave way to the present interglacial period. Radiometric dating, however, indicates that the interglacial occurred about 120,000 years ago.

Based on still earlier interglacial lake beds in Germany and England, similar studies indicate durations of from 16,000 to well over 20,000 years. Yet all the interglacials appear as relatively short episodes in the long, cold-dominated history of the Ice Age.

With such rapid climatic changes, lakes tend to have had a rather short life, just a few tens of thousands of years—short, that is, when we start looking at lakes that existed before the Ice Age. We are then well back in the Tertiary period, when climatic change, as a rule, seems to have been slower, or at any rate, not at all extreme. And we meet with chronologies on another scale entirely.

A fossil lake bed, of the same type as that of the living Lake Valkiajärvi, but from Nevada in the United States, was studied by Michael A. Bell and Thomas R. Haglund. Here too is an alternation between light-colored, diatom-rich summer bands and dark winter bands. The average thickness of an annual varve was less than one-half millimeter (0.347), and the entire sediment layer was about 50 meters thick. Dividing 50,000 by 0.347 gives 144,092; we may round it off and say that about 150,000 years are represented. Interestingly, it turned out that this time was long enough for a significant evolutionary change to occur

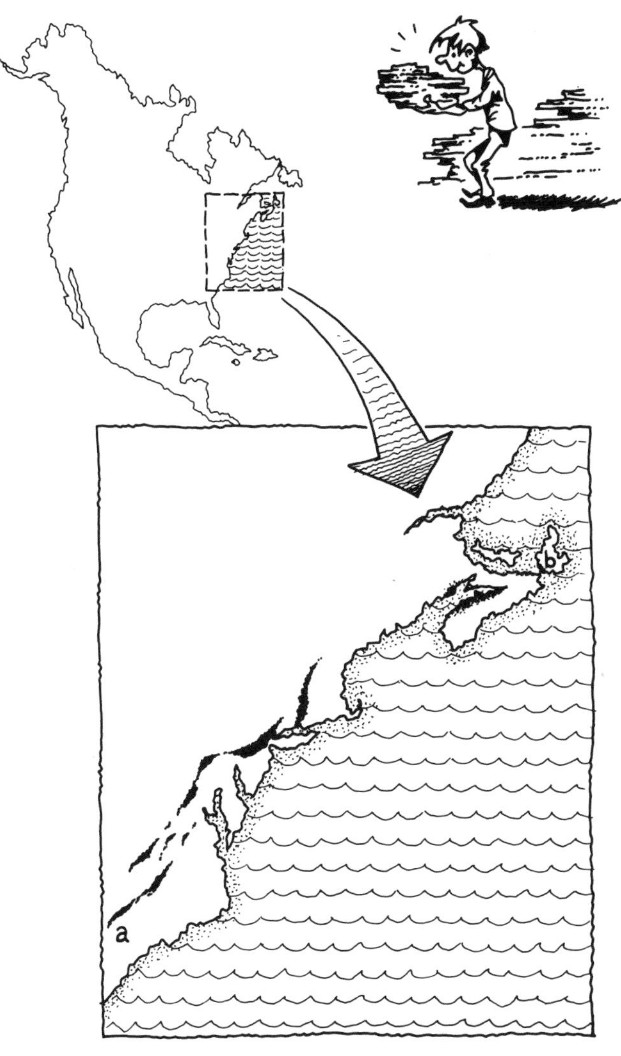

The strata in the 200-million-year-old rift valley follow the east coast of North America from North Carolina in the south (a) to Nova Scotia in the north (b). (Source: Paul E. Olsen)

in a fish that lived in the lake—a species of the stickleback genus *Gasterosteus*.

This lake in Nevada dates from the Miocene epoch, but of course existed only during a fraction of the epoch. In Europe too lake beds with annual varves are known from the Miocene, for instance at Öhningen by Lake Constance; or from the Oligocene, such as Rott in the Siebengebirge of western Germany; or from the Eocene, such as Geiseltal in East Germany and Messel in West Germany. In all these, annual varves prove the lakes to have existed for hundreds of thousands of years.

Still, there are longer sequences. During the Eocene epoch, a lake basin formed in North America, covering large parts of what are now the states of Colorado, Wyoming, and Utah. The area had previously been traversed by rivers flowing east, but in connection with the rise of the Rocky Mountains a barrier was formed to the east. Thus arose Fossil Lake, and waxed greater and greater, finally to cover an area of about 13,000 square kilometers, with a depth of up to 100 meters. It was thus about half the size of Lake Erie, and twice that of Great Salt Lake.

Judging from the sediments that lie beneath and above those of Fossil Lake, it existed for about one-third of the Eocene epoch. The lake-bottom sediments show a fine lamination with alternating light, lime-rich bands and dark ones containing a great deal of organic matter. There is also rich fossil flora and fauna (especially fish and insects). The flora indicate a climate of a subtropical type with two annual rainy seasons, and if the varves are interpreted on that basis (two dark bands to a year), it can be seen that the entire pile of sediments was formed during a time period of 6.5 million years.

According to radiometric dating, the Eocene epoch started about 55 million years ago, and ended about 35 million years ago. Its total duration would then be about 20 million years. One-third of that is 6.7 million years, or very close to the number of years that can be counted in the lake-bottom silts of Fossil Lake. The agreement is almost too good to be true, but there you are. Radiometric dating is supported by the chronology based on annual varves.

The longest consecutive sequence of annual varves that I have happened upon, in a far from systematic search, comes from eastern North America and constitutes about 40 million years. It dates from the later part of the Triassic period and the early part of the Jurassic period, and so has an age of about 220–180 million years before the present. My authority is Paul E. Olsen.

The geography of the earth at the beginning of the Triassic period was very unlike that of the present day. All continents were then collected into a single supercontinent, called Pangaea ("all land"). Then began the birth of the Atlantic: a great rift valley running in a northeast–southwest direction started to form. It has its counterpart in the present-day Rift Valley of East Africa, which also marks the place of a future ocean. The Triassic valley, a string of at least thirteen elongated basins, extended from Nova Scotia to North Carolina. In the basins, very fine-grained lake sediments were deposited (forming the so-called Newark Supergroup) with annual laminae, totaling some 40 million. Parts of the rift valley are now covered by the sea, but the southern part is dry, as are patches further north.

The sediments preserve a record of climatic changes, especially an alternation between dry periods, when the water was low, and times of high precipitation and high waterlines. During the former, annual varves tend to be very thin, drying cracks are formed, and there are numerous footprints of reptiles, in some cases so perfectly preserved that you can count the scales on the soles. At high water, the drying cracks are absent, and the sediments contain a great amount of organic matter, especially fish remains.

As it happens now, these climatic changes are cyclical: they tend to return at regular intervals. Their periodicity is complicated, however, because it is a combination of several cycles differing in length. (Each cycle represents a sequence from low through high to low water.) The most important periods are about 25,000, 44,000, 100,000, and 400,000 years in length, respectively. These are figures that cause the geochronologist to smile in recognition. They have to do with changes in the rotation of the earth around the sun, known from astronomical calculations, and have turned out to drive the climatic changes during the Ice Age as well—that is, during the last two million years or so of earth history. Now the analysis of the Newark Supergroup sediments proves that the same factors affected climates as long as 200 million years ago.

And so the geological time scale, originally dated by radiometry, is corroborated by two additional and completely independent methods of study: from analysis of annual varves, and from astronomical observations.

CHAPTER SEVEN

Two-Dimensional Animals

Einstein's universe is simultaneously finite and without end. To explain this paradox it is usual to make use of the following image. Suppose that instead of having three dimensions (length, width, and height), we had only two —length and width. We would then be completely flat, like figures cut from a piece of paper, and would be utterly incapable of imagining a third dimension. Still, if we were moving across the surface of the earth (leaving mountains or seas out of account), we could crawl any distance without meeting barriers or limits; and if we continued to move in the same direction we would finally get back to the place where we started. Similarly, the Einsteinian space curves in time (the fourth dimension), and a journey in just one direction would also end up at its starting point.

A two-dimensional being, of course, is out of the question. It is certain that such a thing could not possibly exist. Still, it seems that once upon a time, life did produce

something along those lines—here on our own earth. It is one of the strangest discoveries that paleontology has made. Once again we see that the "possible," what existed at one time, exceeds our imagination.

The story begins in Australia in 1947, when some odd-looking fossils were found in the state of South Australia,

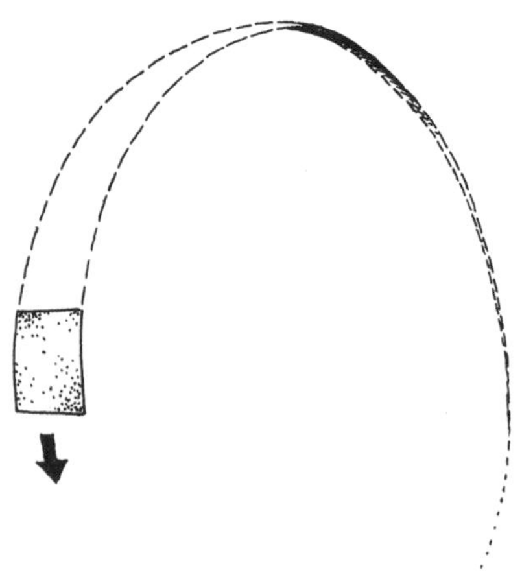

more particularly in the Ediacara Mountains. They are part of the Flinders chain, which runs in a north-south direction north of Adelaide and east of the Torrens Sea. Later other layers were uncovered containing similar fossils in no less than eleven places in the Flinders chain. On a much higher level in the layer sequence, sediments were found with fossils dating from the lower Cambrium period, which of course, shows that Ediacara animal life existed at a much earlier time.

What is so strange about all this? The fact is, the Cambrium is the earliest of eleven periods in all that have provided us with a rich selection of fossils of multicellular organisms, mostly various types of shellfish and other animals with shell-like coverings. The periods have been combined in the "Phanerozoic aeon," a term coined from two Greek words meaning "visible" and "life." (Terms like "period" and "aeon" have sharply delimited meanings in geology.) The geologic history of the earth may be divided into two parts: the Phanerozoic aeon (with a rich harvest of fossils), which began approximately 570 million years ago, and the earlier and much longer still Precambrian period, which reaches back to the birth of our planet—close to 5,000 million years ago.

"Visible life"—this summarizes the contrast between the Precambrian layers, which appeared to be completely lacking in fossils, and the time from Cambrium down to the present, with its abundant life forms. Already Darwin regarded this contrast as a difficult problem. But modern research has changed the picture completely.

Pioneer work by, for instance, the Soviet scientist Timofeev and the American Barghoorn has shown that many of the Precambrian sediments actually preserve the remains of a luxuriant world of microscopic organisms, unicellular beings. Precambrian paleontology is now

AEON	ERA	PERIOD	EPOCH	MILLIONS OF YEARS
PHANEROZOIC	CENOZOIC	QUARTERNARY	HOLOCENE	0
			PLEISTOCENE	0,01
		TERTIARY	PLIOCENE	1,6
			MIOCENE	5
			OLIGOCENE	25
			EOCENE	35
			PALEOCENE	55
	MESOZOIC	CRETACEOUS		65
		JURASSIC		140
		TRIASSIC		200
	PALEOZOIC	PERMIAN		230
		CARBONIFEROUS		280
		DEVONIAN		345
		SILURIAN		395
		ORDOVICIAN		435
		CAMBRIUM		500
	PRECAMBRIAN			570
				4500-5000

Geologic time is divided into *eras* that in turn are divided into *periods;* the periods are divided into *epochs*. The three eras are combined in the Phanerozoic *aeon*. The divisions within Precambrian time are disputed.

undergoing rapid growth, even in Scandinavia, where Precambrian sediments occur in many places. The oldest confirmed discoveries of life forms have been found in layers that are at least 3,500 million years old; that long has life existed on this earth of ours!

As the theory of evolution might lead us to expect, all the early fossil remains belonged to organisms at the very lowest, most elementary level, corresponding to present-day bacteria and blue-green algae, that is, one-celled organisms without nuclei. They are called procaryotes. One-celled organisms with nuclei, or eucaryotes, are first identified with confidence at the level 1,500 million years before the present. Toward the end of the Precambrium you find the first signs of multicelled organisms (for example, filiform, or thread-shaped, algae).

But where does Ediacara belong in this story? Since 1947 Ediacara animals have been discovered all over—in Australia, Namibia, southern China, Siberia, the European part of the Soviet Union, Scandinavia, England, Newfoundland. (A summary by Cloud and Glaessner was published in *Science* on August 27, 1982.) Thus, this primitive type of fauna was spread all over the earth toward the end of the Precambrian period.

What kinds of animal, then, lived during the Ediacara period? Finds from Ediacara lie in a beach of fine-grained sand; they are assumed to have been animals stranded on the beach at low tide, which died and at the next high tide were covered with sand, which helped to preserve them. Moreover, fossil wave marks—of the same type seen at the water's edge even today—have been preserved in the sand.

In the thirty or so species found in the Ediacara sandstone, scholars have thought they recognized relatives of present-day animals: feather corals, jellyfish, and worms.

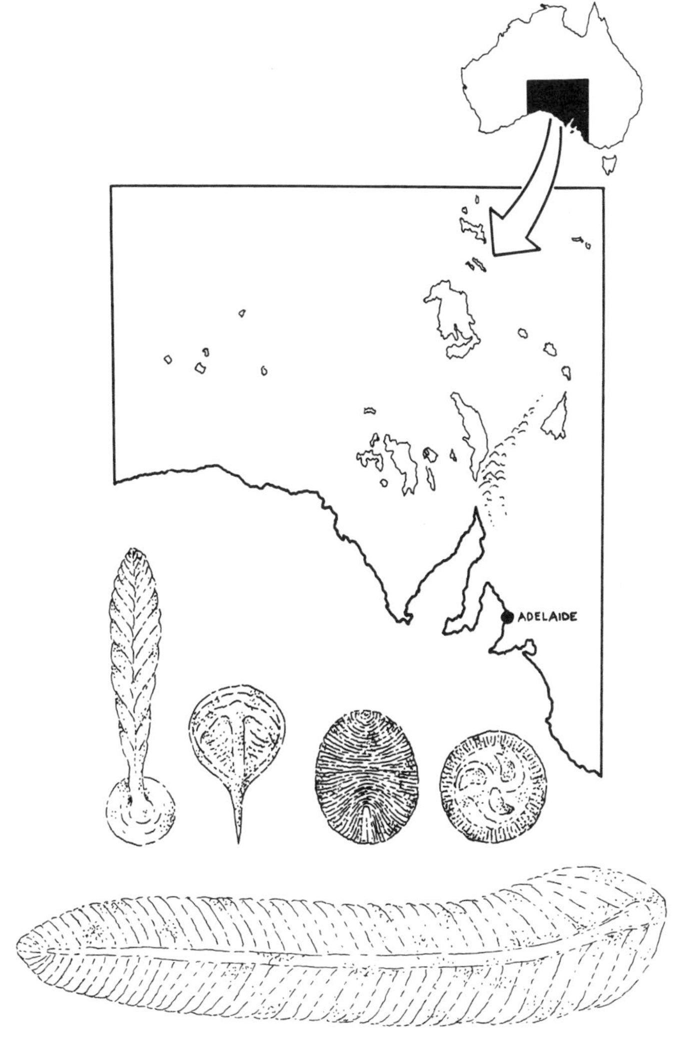

Map of Australia, showing the Flinders chain of mountains (the mountains north of Adelaide). Below: some typical animal fossils from Ediacara. The biggest one, *Spriggina* (bottom), became almost half a meter long.

In other words, it was thought that Ediacara represented an early stage in the development of the multicelled animals as we know them. Now, however, the German paleontologist Adolf Seilacher—one of the leading scientists in his special area—has found that the creatures from Ediacara are not primitive forms of later animals at all but in a way represent a world all to itself, a form of life that no longer exists on earth.

For instance, if we look at the metabolism of a modern vertebrate we find that it has an intestinal canal in which its food is digested, a blood circulatory system that transports oxygen and nourishment, lungs that are filled with oxygen and give off nitrogen and carbon dioxide, and so on: a system of inner pipes and tubes. Similar arrangements are part of the invertebrates. All this is missing in the Ediacara animals. Instead they are flat as pancakes

Flatworm (tapeworm)

(resembling a padded quilt)—even the very largest ones, which could be as long as one meter. Thus, metabolism in their case—the absorption of food and oxygen, the giving off of waste products—took place directly through the skin; of course this works only if the distance between the animal's interior and its exterior is short throughout its body. (Present-day flatworms, among them intestinal parasites like tapeworms, function in about the same way, but the Ediacara animals were not flatworms in the sense that they were related to the flatworms of today.) In fact, you might say that the Ediacara fauna were two-dimensional animals—they had length and width but no thickness! An original but inherently logical solution to the problem of large stature without having to develop a system of internal organs.

The Ediacara fauna, as Seilacher sees them, may thus have been nature's first "attempt" to produce multicellular animals on a larger scale. The experiment was unsuccessful in the long run. Nevertheless, the "two-dimensional" beings managed for quite some time, probably about 40 million years. After that the animals from Ediacara were

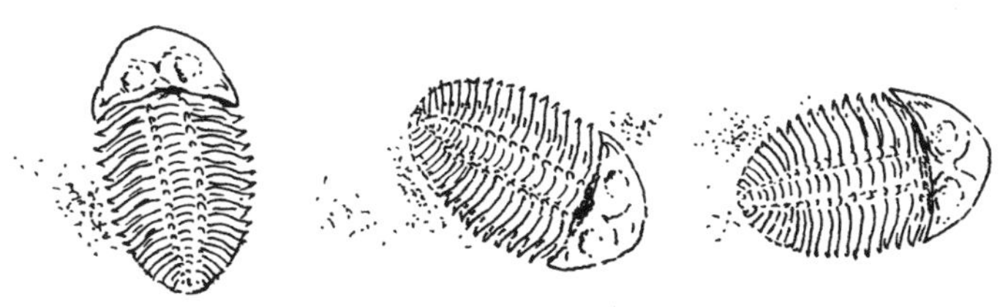

Trilobites

to be subsumed in a pattern that was repeated time and time again in geologic times. For it has been shown that the forms of life that are dominant during one time period usually die out without producing the dominant forms of life in the next period. The trilobites (a kind of insect similar to wood lice), which were dominant in the Cambrian era, died out long ago, and their role has now been taken over by the crustaceans, which are not descended from them. The dinosaurs were the most majestic forms of life during the earth's "middle age," through more than a hundred million years. Still, they died out and their sway over the animal world was transferred to the mammals, which do not derive from them.

It was once believed that the typical lower Cambrium life, with its wealth of trilobites, followed directly upon the Ediacara period. Recent research has shown that an earlier, hitherto unknown time period of approximately 25 million years separated the Ediacara from the lower Cambrium. The period has been called Tommotium. Typical of the Tommotium, in which trilobites are not present (with the exception of fossil tracks that resemble those of the trilobites), are small mollusc-like shells and different types of fragment, which have just now begun to be examined scientifically. In any case, it looks as if we have here the first traces of "three-dimensional" animals of a more modern type.

The transition from Precambrium to Cambrium is thus a much more complicated story than was earlier surmised. The typical Ediacara animals might represent an extinct form of life. But it is likely instead that the faint imprints of worms, which also may be seen in the Ediacara layers, are the first evidence of what the future would bring.

But the last word has not been said in this story. Many scientists maintain, unlike Seilacher, that the Ediacara fauna includes earlier forerunners of corals, jellyfish, ringed worms, and arthropods. And in a fairly recent issue of the journal *Nature,* November 21, 1985, the Australian paleontologist Ian Dyson shows that the Ediacara period may have begun earlier than was believed, and thus must have lasted longer than the 40 million years thought before now.

The earliest fossils of Ediacara animals in the Flinders

chain follow immediately after a layer that was formed during one of the glacial periods. Similar situations, with Ediacara animals covering extremely old moraines, may also be seen in Sweden, Eastern Europe, Newfoundland, North Carolina, China, and Southwest Africa. Did the worldwide glacial period have anything to do with the conditions that led to the creation of this unique fauna?

Like all valuable research, the studies carried out on the Ediacara period have opened up new horizons, posted new questions and new challenges.

CHAPTER EIGHT

The Spoor

Two small figures are moving across the plain. Already the sun is high in the sky and is warming their backs. At the horizon over their right shoulders they can see, through mist and rising vapors, the outline of the volcano—Sadiman—now in repose after its eruption the day before.

They had not been within range of the hot ash cloud, and in the night it had been raining, so that the ash now forms a soft, moist layer of muck on the ground. Marks of the raindrops are still to be seen, as well as the trackways of animals.

The two bipeds are naked. The one who walks in front is a male: he is taller and has big feet. The female, walking slantwise behind him, is slight and graceful; her feet are like a child's. They move on without haste, heading north. In the distance flocks of horses and antelopes can be seen, and a couple of big elephant-like animals still farther away waver, mirage-like, in the shimmering air.

The female stops suddenly. In the corner of her eye she has glimpsed movement. Is there something hiding in the little spinney over to the left? She turns left and narrowly

observes the trees, raising her left hand to shade her eyes: anything dangerous? The great sabertooths hunt in the daytime. . . . No, she is reassured, there is nothing there. She resumes walking in a northerly direction, behind the male, who has continued, undeterred, on his way.

The sun is hot, and the moist ground is steaming. As the day goes on the ash dries out, and when night falls it has become dry and hard as stone. The rain that night makes no impression on it.

Later, Sadiman wakes to life again and spews out a new cloud of ash that forms another layer on the ground and, in turn, is transformed by water and sunshine into stone. And in the following weeks, this is repeated again and again, until the forces of Sadiman are spent for the time being.

This happened some 3,600,000 years ago; and those two walkers on the plain, walkers in a northerly direction, male and female, belonged to the genus that we call *Australopithecus*. We would hardly call them human beings, although they were upright and moved just as surely on two legs as we ourselves. They were protohumans, maybe our own direct ancestors, or at any rate very close to our ancestry.

The tracks are probably all that is left of this particular couple, and they tell the story outlined above. However, Sadiman—the volcano is still in existence—has erupted many times since then, and many other representatives of *Australopithecus* have left their tracks in the ash. Also, there are fossil skulls and bones of *Australopithecus* in the same deposit. In addition, there are many more tracks and bones of animals that lived together with the protohumans at Laetoli; all belong to species that have long since become extinct. The trackways from Laetoli, Tanzania, show that the protohumans walked in the same manner as we do.

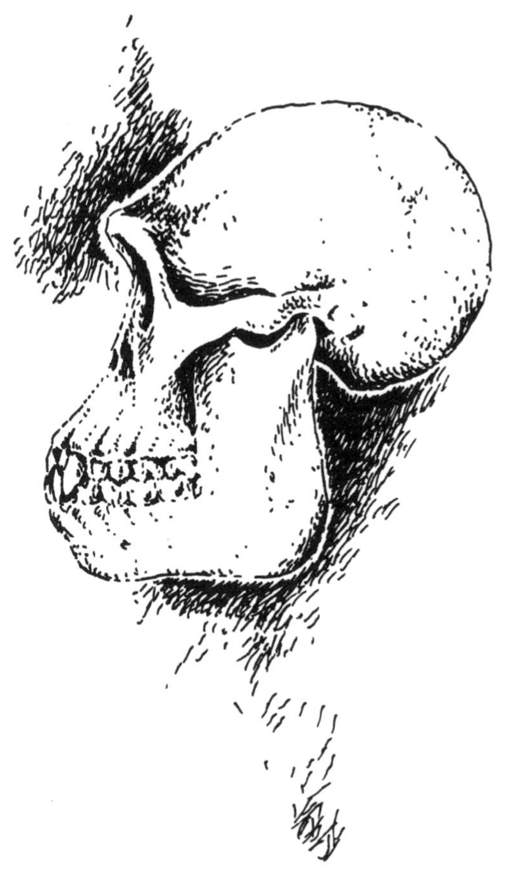

Australopithecus

Footprints would seem to be so perishable that their preservation must be extremely rare. Yet such prints are among the most common fossils, and so there has arisen an entire science devoted to their study: ichnology (from the Greek *ichnos,* footprint). It is not, after all, so surprising. A four-footed animal consists, among other things, of a limited number of bones, which also have a limited chance of becoming fossils; but during its lifetime, the animal produces thousands or millions of tracks—so that even though the statistical probability for preservation of any one is very small, the sheer number is so great that the chance is slightly better than zero. Against all odds, somebody will win the sweepstake, or the Lotto ticket, in the long run.

So the hominids of Laetoli were bipedal. (Hominid means a member of the human family, the Hominidae; we too are hominids.) And this proves that ancestors of humanity walked bipedally 3.6 million years ago. It was the final link in the chain of evidence that had already been forged by study of their anatomy; the footprints proved the anatomists right.

Let us look at an analogous case. It has long been known that the first tetrapods—or four-footed animals—which were amphibians, were derived from the lobe-finned fishes called rhipidistians. The rhipidistians (which existed in the Devonian period) possessed muscular paired fins with skeletons of a kind that could well have evolved into the short limbs of the first amphibians. It was thought that the rhipidistians were able to crawl up onto dry land, for instance in order to find their way to another stream if the one they lived in dried out. It is not too uncommon to find fish behaving in this manner; eels are known to do it, and the present-day mudpuppy can even climb trees. The swim bladder of some fish may function as a lung. But did the rhipidistians actually move overland?

The answer comes from the Orkney Islands. Sandstone from the middle Devonian—some 375 million years before the present—does actually carry the tracks of a fish that wriggled its way on the ground with the help of its powerful fins. This is the oldest known spoor of a vertebrate on dry land.

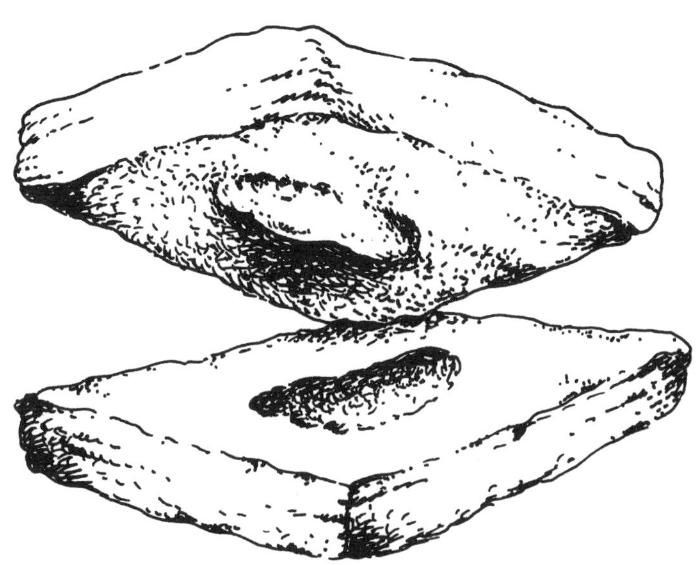

The earliest-known amphibians are late Devonian. Their skeletons have been found in Greenland as well as in Australia. In these animals, the rhipidistian fins have already evolved into five-toed tetrapod limbs (the difference is not so great as might be thought), but they retain much of the piscine inheritance in their anatomy: the tail fin is still there, and so is a vestige of the gill cover. And once more, the footprints clinch the matter. In Australia, three trackways were found in upper Devonian sediments, and in Brazil there is one footprint—just one, but unmistakably of a tetrapod—from about the transition between the middle and the upper Devonian. Thus, the Brazilian print is the oldest known evidence for the presence of tetrapods: some 365 million years ago.

Track sequences made by amphibians of this kind—built essentially like modern salamanders—show the imprint of the entire palm and sole, which suggests that the movements were slow. The digits tend to be oriented outward. The prints of the left and right feet are well separated, showing that the limbs jutted out sideways from the body. The descendants of the amphibians, the reptiles, inherited this type of gait, but in many reptiles the legs tended to swing in beneath the body, which made walking and running possible. The prints of the left and right feet are then close to the midline of the trail. Such changes can be observed in reptile footprints in Mesozoic sediments.

Moist volcanic ash, like that of Laetoli, is almost ideal for taking good imprints, but it is a rather unusual setting. Even here, of course, prints tend to deteriorate in quality the dryer the ash when it takes them. In general the best chances are created on a shore where the water has recently ebbed. A fine-grained sediment will retain moisture for some time and take detailed imprints that may be

preserved if the sediment dries, hardens, and is buried by new sediment at the next flood or tide.

It might be thought that fossil footprints would be easy to counterfeit, but such is not the case. Attempts have often been made. During Depression times in the 1930s, fabricated "human" footprints were peddled by jobless people at Paluxy Creek in Texas; they had been cut out in the same Cretaceous rock that contained many dinosaur tracks. Since then pseudoscientists wanting to "prove" that human beings were contemporary with the dinosaurs searched diligently for human prints in the Cretaceous deposits at Paluxy, naturally without any success whatever. What was touted as human prints has turned out to be a mixed bag of counterfeits, blurred dinosaur prints, and uneven surfaces in general. (There is a detailed study by D. H. Milne and S. D. Schafersman in *Journal of Geological Education,* 1983, pp. 111–23.) A real track shows details that are impossible to counterfeit, such as the compaction of the sediment under the foot. Even to produce an impeccably "natural" shape requires sophisticated anatomical knowledge. And so the most elaborate fabrications are really the simplest to fault—as, for instance, the tracks of the late and unlamented "Sasquatch," California's answer to the Snowman.

It is an accident of history that the Creationist pseudoscientists harbor imaginings about Paluxy Creek itself; this was the first well-known site of dinosaur tracks. There are innumerable others, and some of them are much richer in finds, giving dramatic snapshots of the past. One of the most famous, however, comes indeed from Paluxy Creek and is a well-known exhibit at the American Museum of

Natural History in New York. There we see the tracks of a gigantic quadrupedal sauropod. The track sequences show that the legs were pillar-like and carried the body in the same way as the legs of an elephant. So this immediately renders false the lizard-like stance advocated by certain earlier paleontologists around the turn of the century. It also shows that the giant sauropods were fully able to move about on land and that they were not dependent on a watery environment, as had long been thought.

There we see also the footprints of a big theropod, or bipedal predatory dinosaur, which has been walking in the tracks of the sauropod. It looks like a hunting scene, but there is no way of telling how it ended.

A special problem arises whenever connected series of prints cover very large surfaces—it becomes enormously costly to collect the material and find space for it in museums. An Australian find from the middle Cretaceous period is a good example. There they had to deal with a surface of over 200 square meters, covered with more than 3,000 prints. The find is located in Queensland, between the towns Winton and Jundah. Richard A. Thulborn and Mary Wade have carried out a precise analysis of the whole impressive collection of prints and have pieced together the following account of what was enacted more than 100 million years ago.

The setting is the northeastern shore of a lake, which had shrunk down to normal size following a flash flood that had deposited a thick layer of clay; a promontory about 60 meters long and 20 meters wide juts out toward the southwest. Numerous small dinosaurs, probably several hundred, have made their way to the western shore of the promontory to slake their thirst. They are all bipedal but belong to two completely different types of dinosaur. Something less than half of them are herbivores,

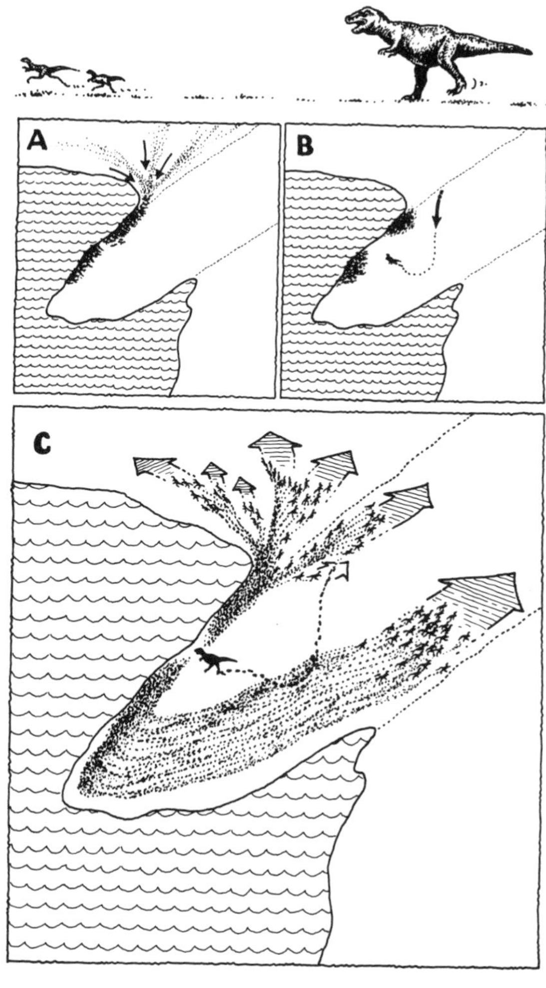

Reconstruction of the course of events at the Australian site where the many footprints were found. **A**—a big flock of small dinosaurs gathers by the shore. **B**—the large predatory dinosaur enters the scene. **C**—mass flight by the small dinosaurs; many run in the tracks made by the big dinosaur. (Source: R. A. Thulborn and M. Wade)

and the rest are carnivores. Both types leave behind imprints with three toes that are similar to those of birds; but the prints of the herbivores are broad with short toes, while the carnivores reveal elongated toes with sharp claws. At times, when a foot has sunk deeply into the clay, the imprint of a much shorter fourth toe (actually the first one, corresponding to our big toe) becomes visible. The dinosaurs vary in size—the smallest are no bigger than hens, while the big ones rival an ostrich. Perhaps there are several species; perhaps the group included both adults and the very young.

The feet of a small herbivorous dinosaur (left) and a small carnivorous dinosaur (right). The reconstructions are based on fossil bones and tracks. (Source: R. A. Thulborn and M. Wade)

Then, however, something happens that makes both flocks take note. Toward the northeast appears a gigantic figure, a huge bipedal meat-eating dinosaur. Perhaps it emerges suddenly out of the forest. It is a real giant; its feet leave imprints that measure over half a meter in length; the length of its leg from the hip joint down is more than two and a half meters; the monster may be all of ten meters long. Its huge mouth is filled with teeth like razors.

The predatory reptile at first moves toward the southwest, toward the tip of the promontory, but slows and then turns to the right, directly toward the large flocks on the shore. By now it has left eleven huge impressions behind on the soon-to-be-deserted surface, eleven earth-shaking paces covering a full 17 meters, and the three last show a turn to the right.

Panic breaks out among the small dinosaurs on the shore when the ravening giant towers over them. They flee in all directions, but there is only one way to safety: the way from the promontory to the mainland, which is the same route taken by the giant marauder. Herded together, they rush toward the northeast, but they are packed too closely together to get up much speed—perhaps they are also out of breath. The imprints of the giant's feet make pitfalls in their way as they push and shove one another and stumble, trying to get out of its way. Many do not succeed; they sink down into the footprints and jump up again.

The enemy does not pursue them. It probably caught one of the small dinosaurs and is now busy eating its prey.

At least 150 dinosaurs rushed across the surface, which is approximately thirteen meters wide and has been preserved. Others, perhaps just as many, may have jumped to either side. They have all long turned to dust; only the footprints remain. But they still create an image that we can see, its contours as sharply etched as if we had watched it beneath the burning sun of the Cretaceous, an image of the great panic that occurred a hundred million years ago.

Maybe someone will be inspired to write them an ode: with the "noble six hundred" we could have, perhaps, "the terrified hundred and fifty."

The huge task of collecting the prints the Australians solved by casting the entire surface, footprints and all, in fiberglass—much easier to keep and preserve in a museum!

The Australian find suggests that certain dinosaurs lived in flocks and thus were social animals. We know many similar cases. In Texas the tracks of a herd of 20 giant dinosaurs have been found, all traveling in the same direction. The smallest ones walked in the middle and the biggest ones formed the flanks of the group, which indicates that the adults were protecting their young. In another locality the tracks have shown that bipedal herbivorous dinosaurs acted similarly—here the footprints of 25 individuals were found, probably of the genus *Camptosaurus*. Even predatory dinosaurs seem at times to have hunted in groups. One example comes from Massachusetts, where 19 individuals of the genus *Eubrontes* would move forward in a wide front (no prints overlap others). *Eubrontes* was a relatively small predator. Interestingly, this find dates from the Triassic period, that is, before the heyday of the big dinosaurs in the Jurassic and Cretaceous periods. So social behavior occurred at an early stage in the long history of the dinosaurs.

Finally, I will mention a curious find at Peace River in the chalk deposits of western Canada. It includes no less than 1,700 dinosaur tracks, plus a series of prints made by a bird and another one by a turtle. Most of the dinosaur prints were made by a flock of the very large bipedal herbivores called hadrosaurs, also known as duckbilled dinosaurs. The adults walked next to one another in a wide front, the young ones trailing behind (they often trod in the tracks made by the adults). At one spot you can see that eleven animals suddenly changed course from south to east. The turn was made so abruptly that some of them had to jump aside so as not to collide with others. Here is also evidence of small predatory dinosaurs traveling in groups, while the larger species seem to have been solitary travelers, or perhaps walked along in pairs.

These are the kinds of find that have forced scientists to revise many of their older ideas about dinosaurs. It was once thought that the quadrupedal giant sauropods lived in the water; the largest of them weighed 80 tons or more, and it was thought that they were just too heavy to walk on dry land. But the footprints show they made out quite well and contribute to the making of an entirely new picture of these fantastic beings. Since their dental armament was strikingly feeble, it was thought that they lived on various kinds of easily chewed aquatic plants. Another possibility was not even considered, namely that they may have had a muscular gizzard, in which the food was worked over with the aid of gizzard stones, after which it could receive follow-up treatment by intestinal bacteria. The crocodiles, next to the birds the closest relatives of the dinosaurs, have exactly the same kind of digestive equipment, which in many ways is more effective than the mammals' method of chewing their food (one reason being that worn-out stones, unlike worn-out teeth, can be exchanged for new ones.) A skeleton of a precursor of the giant sauropods, the *Massospondylus,* found in Zimbabwe, still had a small collection of such gizzard stones inside its ribcage. The closest place where this type of stone could readily be found was up to 25 kilometers away from the find; so we see that *Massospondylus* was not satisfied with just any kind of stone! Remains of true giant sauropods have indicated that they too had stones in their gizzards. It is now supposed that these long-necked dinosaurs fed on the leaves and branches of trees in about the same way as our present-day giraffes.

The evidence of a developed social behavior pattern is confirmed by other finds. In the state of Montana hadrosaur "nurseries" have been found, with as many as 18 young one or two weeks old. That they remained behind in the nest suggests that they had an inborn social instinct and that their parents were sheltering them after they had hatched. Another example, from Mongolia, is a nest with eggs of a small (2 meters long) quadrupedal dinosaur, belonging to a group reminiscent of the rhinoceros. Right next to the eggs were the remains of a predatory reptile with a crushed skull; it looks as if it was killed by the rhinoceros-like reptile in defense of its eggs. Thus the dinosaurs can be regarded as anything but "evolutionary failures"—a popular misconception. They were actually very advanced animals, in many ways more advanced than most of our present-day species. They dominated the earth for 125 million years, and so were the most successful animals that have ever existed.

The perseverance of the ichnologists, the specialists studying animal tracks and footprints, has made it possible for these and many other animals long extinct to live again and move around under our eyes. And for that we can be thankful—for the fact that what may seem most ephemeral, a trail or a footprint, may be preserved through aeons of time.

CHAPTER NINE

The Innocent Assassins

The Zoological Museum of Helsinki University, established in 1985 under the direction of Chief Curator Eirik Granqvist and myself, features a lifesize model of the sabertooth tiger *Homotherium serum.* The name is not derived from the Latin *homo,* meaning "man," but from the Greek *homo-,* which means "similar to." The full meaning is thus something like "a similar animal in our day"; the one who named it was evidently comparing it to the much earlier sabertooth tigers. The word "tiger" is not to be understood literally; the animal was no more closely related to a tiger than a common domestic cat. We are talking about an extinct feline animal about the same size as the modern Siberian tiger but of considerably different body proportions and with tremendously elongated fangs, shaped like scimitars, in its upper jaw. Its head was longer and much narrower than our present-day tiger's. Its neck was longer and thicker than the tiger's, and its forelegs were longer than its hind legs, so that its back sloped to the rear. The long forelegs were primarily due to the great length and size of their lower part, of the wrists and the metacarpal bones. Its tail was short, about the same size as the lynx's tail. A peculiar detail is its claws, which were retractile like a cat's and rather poorly developed. A glance at the reconstruction will immediately show that here we are not dealing with a tiger in the accepted sense of the word but with an entirely different kind of animal (see the frontispiece).

Homotherium serum lived in North America during the glacial period and became extinct approximately 12,000 years ago. A closely related species, *Homotherium latidens,* flourished in Europe about the same time but seems to have become extinct somewhat earlier. The reason we chose to reconstruct the American rather than the European form is the fact that complete skeletons of the former have been found (in a cave in Texas), while we have found only fragmentary bones of the European species. In our reconstructed model we used a skeleton that is on display in the Texas Memorial Museum in Austin, one I have examined myself. You can reconstruct the animal's musculature from the bones that make up the skeleton and thus get an idea of how the animal looked when the bones were covered with flesh. The one thing we know nothing about is the color of its fur. Here there is much room for imagination, and we have chosen to make the animal black, just like its favorite prey, the mammoth, about which I will have more to say later.

Like other sabertooths, *Homotherium* is characterized by enlarged upper canine teeth. In the *Homotherium* they are of the kind usually called scimiter-shaped—only moderately enlarged, flattened when seen from the side, but with sharp serrated dentures in front and back. Another type of sabertooth was the "dirktooth cat," for instance the *Smilodon* ("knifetooth") pictured here, in which the fangs often became quite long. Actually, during the last 30–35 million years of the earth's history a great number of

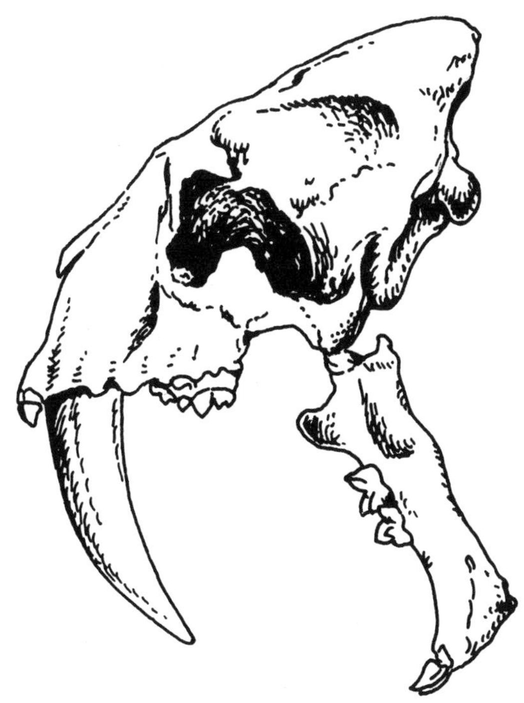

Smilodon

sabertoothed carnivores of various types have appeared, and we know of many intermediary forms between "normal" carnivores and the extreme sabertooth form.

Though I coined the terms "scimitar cat" and "dirktooth cat" myself, I would like to point out that actually their fangs are not much like scimitars or straight-edge daggers. There is a curved Arabian dagger called the *jambiya* whose shape is much closer to the shape of the teeth in the dirktooth.

It was once believed that all the sabertooths made up a

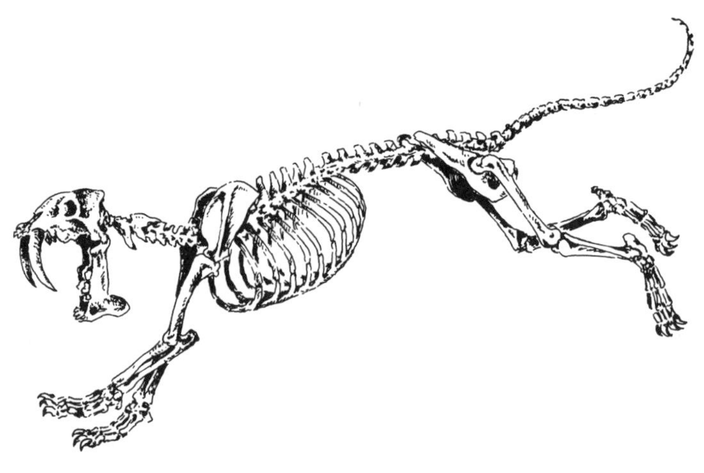

Skeleton of the giant nimravid *Barbourofelis fricki*

single group of closely related species. More recent research, however, has shown that there are two different families of these carnivores, not at all closely related to each other. *Homotherium* and *Smilodon* belonged to the cat family, Felidae, just like all the cats of the present day, from domestic cats to lynxes and leopards, lions and tigers. The other family, which became extinct long ago, is known as Nimravidae, and so we may call its members nimravids. The earliest nimravids appeared about 40 million years ago, and the family was still in existence until about 5 million years ago, at which time the last of the nimravids vanished from the face of the earth. The lifespan of the family overlaps that of the genuine cats, since the first Felidae appeared at least 12 million years ago.

Let us look at the (as far as we know) last nimravid, which lived in North America about 5 million years ago. Its scientific name is *Barbourofelis fricki* (its name immortal-

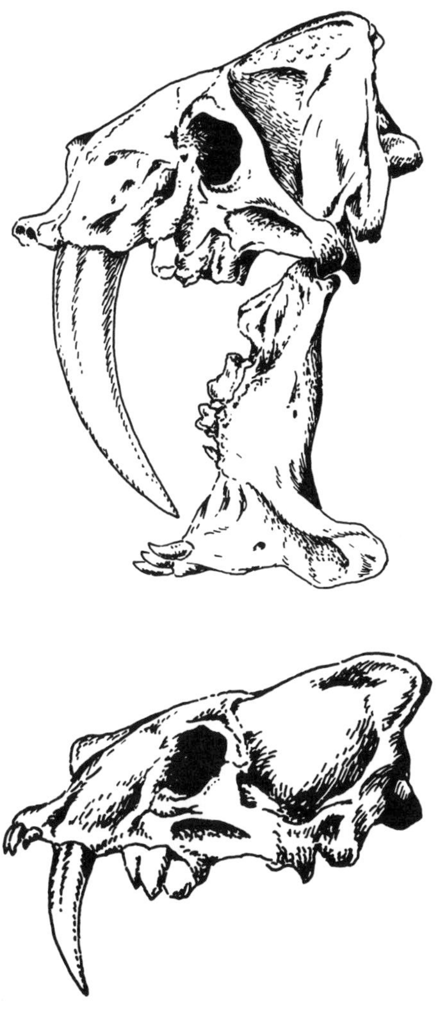

Above: The skull of *Barbourofelis fricki,* which lived in North America 5 million years ago. Below: The skull of its precursor, *Barbourofelis morrisi,* which is 2 million years older.

izes two outstanding paleontologists, Barbour and Frick) and it was the most extreme form of the known sabertooths. Its fangs were unbelievably long. To use them the animal would have to drop its lower jaw along a curve of no less than 115 degrees. Most catlike animals open their mouths at no more than 70 degrees. When the mouth was closed, the sabers were covered by a downward-pointing flange on the chin that functioned like the scabbard of a sword. (Such a flange is found among all sabertooths, with one exception: *Smilodon* was unique in its lack of such a flange.)

Fossils indicate that this species evolved over about 4 million years from a species with short fangs and a less projecting chin. The original species, which lived about 9 million years ago, was about as large as a modern lynx and is called *Barbourofelis whitfordi*. There was an intermediate stage represented by *Barbourofelis morrisi,* which lived about 7 million years ago and was about the size of a puma or cougar. Its fangs and chin both increased considerably in size over earlier species. The development culminated with the *B. fricki,* lion-sized and with incredibly large fangs.

Like the dirk-type predators in general (and in contrast to the scimitars) the *Barbourofelis* was powerfully built. It reminds you of a bear or a gigantic badger more than a cat. Its forepaws were enormously powerful, while the hind legs were comparatively short. Its shinbones were actually shorter than its fangs.

At that stage of its development the *Barbourofelis* genus died out; the nimravids also seem to disappear from our story at that time. In the Old World they had become extinct many millions of years earlier.

Why did *Barbourofelis* die out? To supply an explanation we must first answer another question: why did their teeth grow so large? What is the reason that this kind of dentition has evolved in so many cases?

The first in this evolutionary line is represented by the genus *Hoplophoneus* (the name means "armed murderer" —the invention of names for sabertooths never seems to flag) that lived as early as 30–35 million years ago. This animal, many species of which are known, reached the size of a leopard but was of an entirely different build; it is remotely reminiscent of a badger or wolverine rather than any member of the cat family. That seems to be a common trait among the nimravids, to the extent that their build is indicated by the skeletons found (for *Hoplophoneus* whole skeletons have been unearthed).

At about the same time a considerably more extreme dirktooth, *Eusmilus* ("genuine knife") appeared, and for this one, too, we know several species. While the largest ones approached *Hoplophoneus* in size, there were also species that could be regarded as miniatures: one of them was no bigger than a domestic cat. But it was a formidable cat indeed, which could have had the neighbor's Doberman for breakfast, according to Larry D. Martin, one of the scientists who has done research on this animal. This miniature monster is without a doubt one of the most amazing discoveries ever made in mammalian paleontology, and its way of life is pretty hard to picture. Normally, small predators hunt even smaller prey (cats hunt rats), while it is only among animals of greater size, ranging from lynx to lion, that the prey tends to be bigger than the hunter—the lynx, for example, specializes in hunting deer. But *Eusmilus'* sabers could be effective against even larger prey.

A third group includes the slightly built, scimitar-toothed nimravids *Nimravus Dinictis* ("terror cat"), and other fleet-footed carnivores the size of a puma.

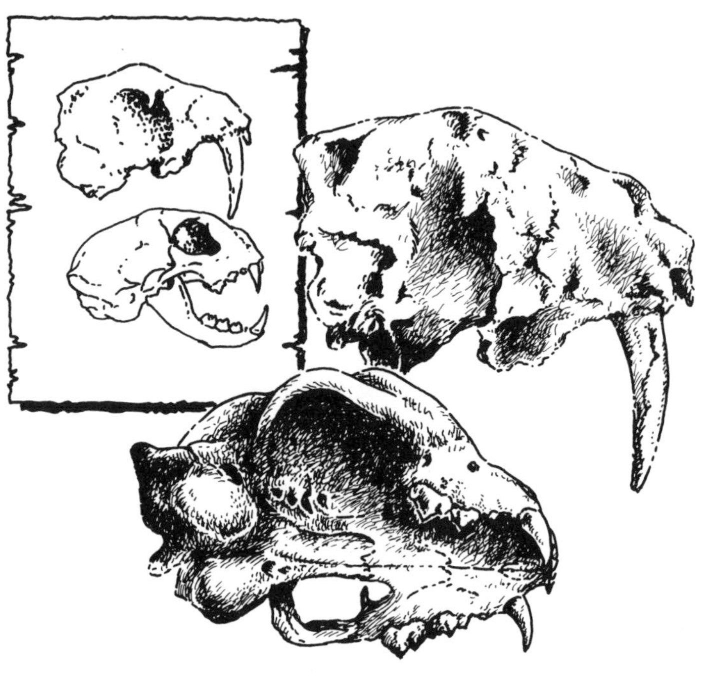

The skull of the mini-sabertooth *Eusmilus,* compared with that of a common domestic cat.

A fourth evolutionary line is represented by *Barbourofelis,* described above, and a fifth one, to this day wrapped in mystery, is the African *Vampyrictis* ("vampire cat"), which lived about 12 million years ago. So far the only remains known are the stiletto-like fang and a molar. (The memory of that discovery is still vivid in my mind's eye, in a light of transfiguration. I unearthed the two teeth while on an expedition in Tunisia, and it took several months to learn for certain what I had discovered.) There are also vestiges of at least one, possibly two, European species of nimravid flourishing 15 to 30 million years before the present. What is so strange about them is that they all seem to have developed completely independent of one another, from nimravids with "normal" dentition.

As I pointed out above, we must regard as genuine members of the cat family both the *Homotherium* of the Ice Age and its precursor before the glacial period, *Machairodus* ("sabertooth"); as well as *Smilodon* and its forerunner *Megantereon* ("big chin": this genus was characterized by a large chin flange, but in the daughter species of *Smilodon* the flange was gradually reduced in size). Thus, we have eight or nine evolutionary lines distributed among nimravids and felines.

But that is not all. Also among what we might call the carnivore prototypes—not closely related to the carnivore proper—were a few similar to the sabertooths, and even the marsupials have brought forth a "marsupial sabertooth," *Thylacosmilus,* which lived in South America about 10 million years ago. Like so many sabertooths it had greatly elongated upper canine teeth and a large chin flange. (Still another "extinct" marsupial sabertooth from an earlier epoch has been discovered recently in South America!) So we find that sabers as weapons of attack have been developed at least ten to twelve times—and we will have to reckon with future discoveries that will add to the total.

How are we to explain this bewildering multiplicity of sabertooths?

The ordinary feline animal attacks its prey by *biting,* in which the canine teeth in both upper and lower jaws puncture the body of the victim. (The front paws are also used in such attacks). In the sabertooth cats the canine teeth of the lower jaw were very small, and the attack occurred mainly by a downward stab with the big sabers. The teeth of the lower jaw probably contributed to the effect of the stab by offering resistance, but in themselves they could not cause any great damage.

The teeth were used in a slashing movement with the stab following an arc that corresponds to the curve of the fangs. It is not hard to show that such a stab could not penetrate deeply into the body of the victim but would run parallel to the surface. So the sabertooth (in its turn, the nimravid) in all probability chose places where big blood vessels lie near the skin, the neck likely being the easiest point of attack. After the teeth had sunk in, the sabertooth could rip backward and tear open two big wounds from which the prey would soon bleed to death. The longer the sabers, the bigger the wound that could be

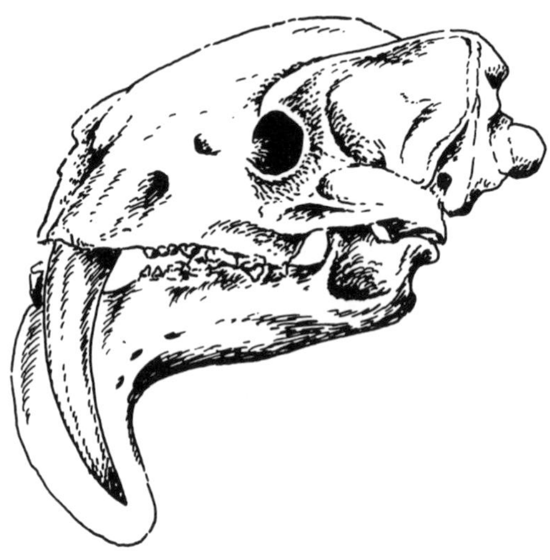

Thylacosmilus

produced, and the more probable that major blood vessels would be torn open and kill the prey. Thus we see that natural selection could lead to an increase in the length of the teeth.

An interesting difference between the sabertooth cats and "normal" felines is that the eyes of the latter are larger. Nocturnal animals generally have larger eyes than animals active during the day. Most of the modern felines hunt their prey in the twilight. We can suppose that the sabertooth's hunting techniques required better lighting—the strike had to be carried out with great precision, one reason being the importance of not hitting a bone and breaking its teeth. That this happenstance was not always avoided is shown by a few *Smilodon* craniums found at Rancho La Brea where you find one and sometimes even two broken fangs (the stumps are usually fairly worn, which shows that the animal survived some time after the accident).

The body build varies quite a bit among the various types of sabertooth. The dirktooths, for instance *Smilodon* and especially *Barbourofelis,* were usually powerfully built, with relatively short legs, and were scarcely swift runners. The scimitars (for instance *Homotherium*) were mostly long-legged and ran very fast. Another difference is the size of their brains. The dirktooths (such as *Smilodon*) have rather small brains—they probably attacked their prey from behind, which doesn't demand too much intelligence. The scimitars had larger brains; their hunting technique (chasing their victims) was probably more demanding. But the techniques vary greatly among the various species, and we can suppose that each species specialized in a particular kind of prey. That is also fairly true of the nimravids; all, however, had one feature in common, namely, their brains were considerably smaller than those in the genuine cat species—but even so, the fleet scimitars had larger brains than the solidly built dirktooths.

Happily, we know what *Homotherium's* favorite prey was. In the Friesenhahn cave in Texas, where mass skeletal remains of scimitar cats have been found (including all ages from newborn cubs to very old animals with worn-out teeth), we have also found what is left of their prey. The find includes hundreds of mammoth milk teeth, which suggests that mammoth calves made up the staple diet of these scimitar cats. (Some milk teeth are also found from the mastodon, a more primitive elephant-like animal, distant relative of elephants and mammoths; but they are comparatively few in number, which might be because the mastodon was less often encountered in this area than the mammoth.)

The find in Texas gives us some indication of why the *Homotherium* died out. It obviously preyed on the mammoth, and since the mammoths (and even the mastodons) became extinct toward the end of the Ice Age, the *Homotherium* lost its food supply.

The *Barbourofelis* died out about 5 million years ago probably also because its quarry became extinct. It is thought that this dirktoothed nimravid hunted mainly rhinoceros and young mastodon. But at that time, all the rhinoceroses died out in America (*Barbourofelis'* home), and even the American mastodons declined sharply in number. (Their death may have been caused by the climate's becoming drier, so that grasslands spread out at the expense of the forests; both rhinoceros and mastodon may have been leaf-eaters rather than grass-eaters). Thus we can suppose that the *Barbourofelis* had no way to sustain itself.

The same fate probably overtook all the various lines of

sabertooth carnivores, which, however, died out at different points in time. They specialized in certain quarry and could no longer manage when an upheaval in the fauna led to the extinction of their quarry. But slowly other species of suitable prey were evolving and at the same time a new group of sabertooths arose to hunt them. Perhaps in a few million years, a new *Homotherium* stock will go on the hunt—if we give it a chance. . . .

And why not? Few, if any, of the extraordinary beings of the distant past can compare with these creatures. The dinosaurs, of course, are unrivaled in their ability to fire our imagination, in their variety and their giant size, but they remain alien to us. The sabertooths seem as close to us as lions and tigers, lynxes and panthers and yet retain the romantic allure of of their superb equipage; they are what Loren Eiseley, poet, philosopher, and paleontologist, has called "The Innocent Assassins." In a poem of that name he tells us of discovering a sabertooth cat "far down in all those cellars of dead time." The sabertooth is as "beautiful as Toledo steel," a masterwork created for the art of killing, formed by the same forces that built mountains from the flat sea bottom of the Cretaceous. Still, it was no ordinary death that was recorded by this find. Instead of slashing into soft flesh the sabertooth had hit a shoulderblade and got stuck there; the tooth was still, so many millions of years after the event, firmly ensconced in the pierced bone. Thus this particular sabertooth met its fate. Even the most ingenious adaptation can sometimes turn into its exact opposite.

As the symbol of innocent savagery, or savage innocence, the sabertooth takes on an archetypal dignity. And so it has become our extreme opposite: the cat, from nature's hand perfectly armed, is the innocent one; humanity, whom nature left unarmed but who has armed itself, is ever guilty and conscience-stricken. Eiseley concludes his poem with the following lines:

We are all atavists and yet sometimes we seem
wrapped in wild innocence like sabertooths, as if we
still might seek
a road unchosen yet, another dream.

He gives us both a lesson and a reminder. The lesson contains the truth that too much specialization may lead to the death of an entire line. The reminder is contained in the still unchosen path, the new dream.

CHAPTER TEN

Argentina's Magnificent Bird

The "biggest in the world" department is alive and well. And not only sensation-hunters and the *Guinness Book of World Records* might be interested in what I'm going to discuss in this chapter. Giant specimens in the animal world are also fascinating to scientists, to whom they represent a number of biological problems. How can you explain animals that grow as large as the whales? How can such enormous beings function, mechanically and physiologically? Such questions are especially apt when we consider airborne creatures, which have to resist the force of gravity in a medium that does not give them much help in staying aloft (a problem not faced by the whales).

There is a general misconception to the effect that the dinosaurs died out because they got too big. However, the dinosaurs lived over a period spanning more than a hundred million years despite their large size. (It is probable that they existed so long *because* they were so large; it

was part of their extraordinary adjustment to the world around them.) On the other hand, we know that untold numbers of microscopic animals have died out in the course of time. So a large body is not necessarily a bad omen for survival, a fact we should note with pleasure, since human beings too are one of the large mammals.

The ostrich is the largest of the now existing birds. With a height from beak to foot of almost 3 meters and a weight of up to 144 kilos (about 317 lbs.), it is tops in its class. Still larger were New Zealand's extinct moa bird and the also extinct Madagascar ostrich, whose weight is estimated to have reached 450 kilos (about 990 lbs.). But they were all flightless birds. No bird that can fly comes even close to such dimensions.

The heaviest bird to be found in Scandinavia—and also the longest one—is the mute swan, whose weight is given

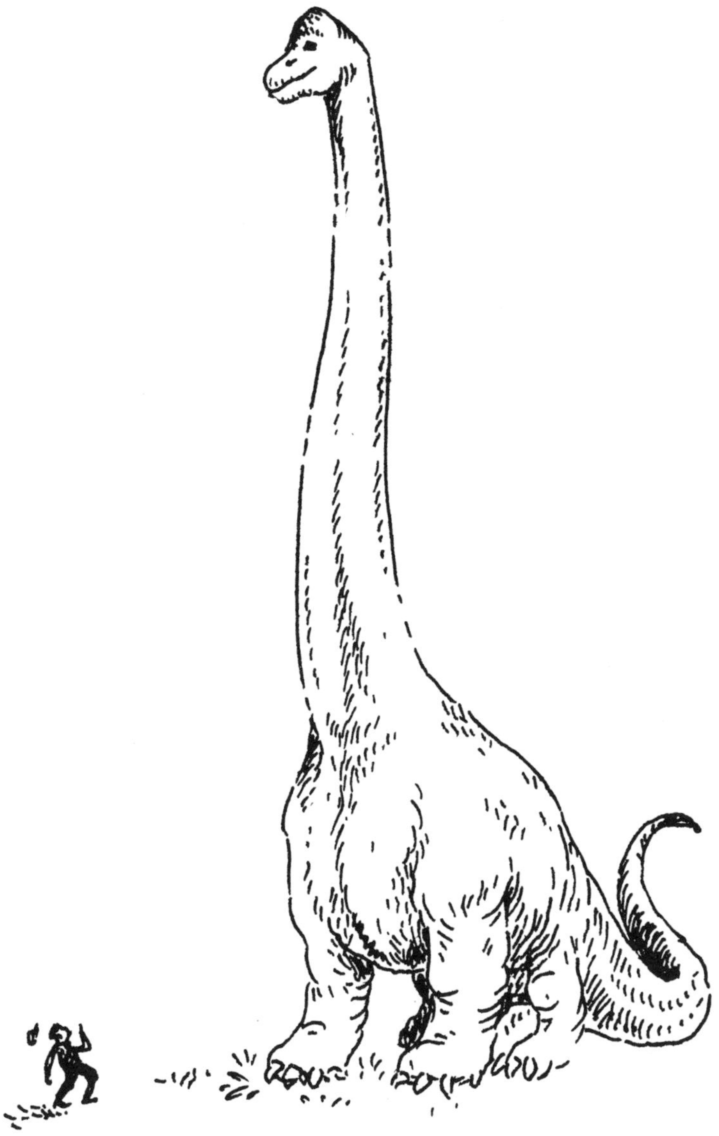

in encyclopedias as 20–23 kilos (44–50 lbs.), with a wingspan of almost 230 centimeters. Its length may be as much as 1.5 meters, but much of the length, of course, is the famous swan's neck.

The largest bird in Scandinavia, in wingspan, is the sea eagle, also called the white-tailed eagle; the span between the wing tips exceeds 2 meters and may sometimes reach 2.5 meters. Many birds of prey in various other areas are still bigger, and that is especially true of the vultures; the biggest of all is the South American condor, with a wingspan of more than 3 meters. The very biggest are certain albatrosses; the record is held by the wandering albatross, *Diomedea exulans,* which can measure up to 3.4 meters between its wing tips.

As late as 1971 a well-known ornithologist, R. W. Storer, could write: "The larger albatrosses, pelicans, storks, swans, condors, turkeys, and cranes must be representa-

tives of the biggest format flying birds can reach." You might suspect that he had to eat those words.

We turn now to the fossil evidence of birds in the distant past. Funnily enough, since the 1950s there has been a veritable competition to see who has the world's record in wing spread. It is all happening, as is only to be expected, in the New World. Let us first look at the situation around 1952.

The biggest bird known until then lived at the end of the Ice Age in the western part of North America. Hundreds of finds in the asphalt layers at Rancho La Brea in Los Angeles provided evidence of the existence there of the gigantic *Teratornis merriami* ("Dr. Merriam's wonderful bird"), described as early as 1909, a huge raptor with a wingspan estimated at between 3.5 and 3.8 meters and a height of 75 centimeters. Its weight was calculated at first at about 23 kilos, but later studies have arrived at 15 kilos as its more probable weight. Sizes and weights on that order are reached among the birds of prey by the vultures only, and so *Teratornis* was regarded as a vulture, feeding on carrion. We will deal with its way of life shortly. In any case, it was written about as the biggest airborne bird,

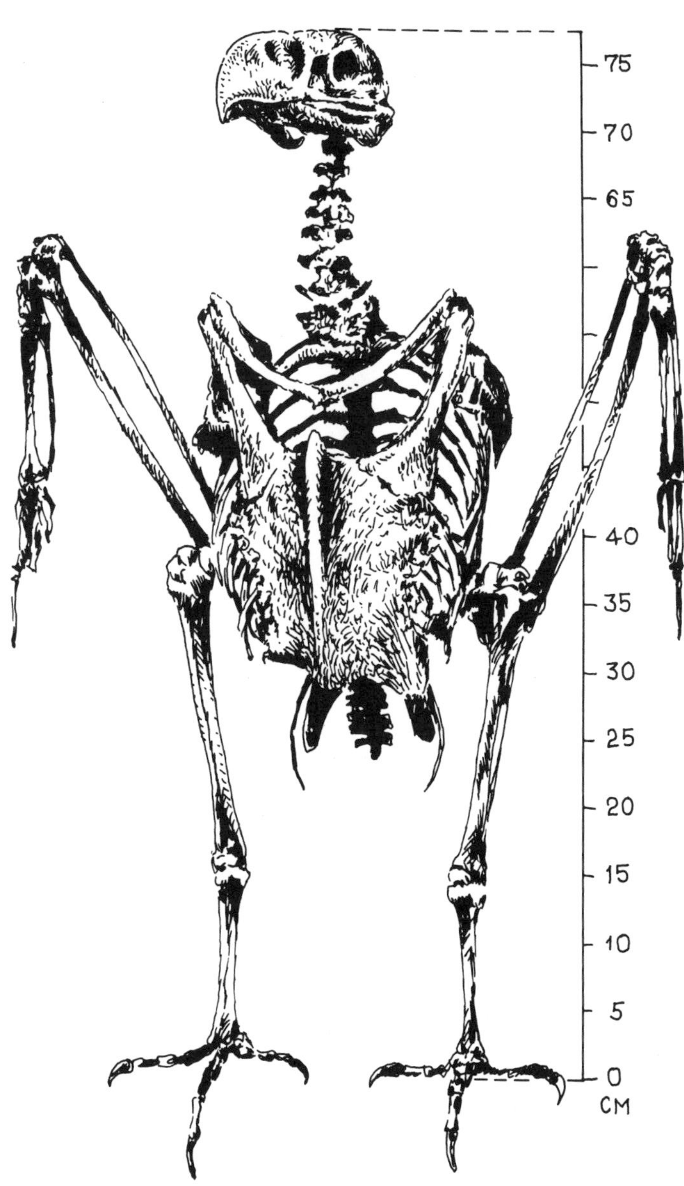

Teratornis merriami

and many people concluded that it was the biggest possible bird that could fly. Its size is so close to the theoretical maximum projected by Storer that in a pinch it may be included within his limit. However, in 1952 the distinguished American paleornithologist Hildegarde Howard announced the discovery of a new species in the *Teratornis* genus, which with an understandable sense of wonder she christened *Teratornis incredibilis*—the "incredible one." Two additional finds involving the bones from this species have been made since then and have been described by Howard. They are all approximately 40 percent bigger than the corresponding bones of *Teratornis merriami,* and Howard estimated its wingspan at about 5 meters.

The "Nevada vulture," as the new species is called, held its world's record unchallenged for five years. Then a contender appeared who at least seemed to verge on a new record.

This time it was a dark horse, if we may use that expression in this connection, namely a bird that belonged to an entirely different order: Pelecaniformes, the pelican-like birds. A highly peculiar subgroup of this order had long been known: a big seabird with an odd-looking set of "pseudo-teeth" in its jaws. All the birds that exist today are, of course, toothless, and this bird's "teeth" were simply pointed tines jutting out from its jaws. In 1957, Hildegarde Howard announced the discovery of a giant member of this group, *Osteodontornis orri.* The exceedingly well-preserved skeleton indicated that its wingspan must have been around 5 meters, thus close to the size of *Teratornis incredibilis.* But in this case the bird was represented by a complete skeleton, while the big *Teratornis incredibilis* was known only on the basis of a few isolated bones.

The "sawtooth" (the family name means "bone-tooth

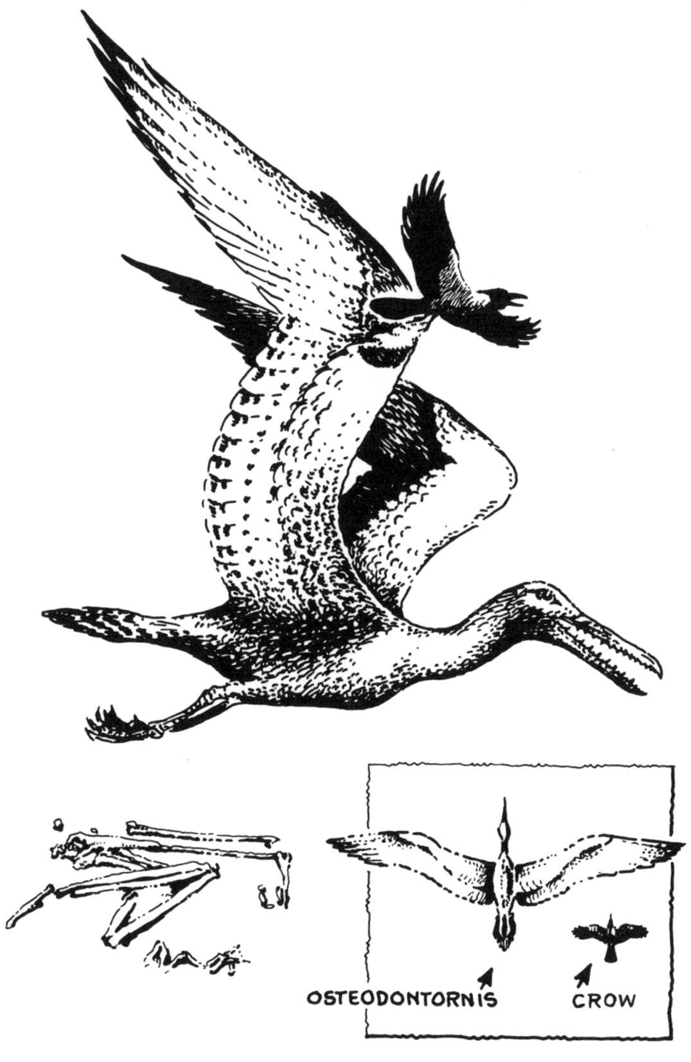

The giant bird *Osteodontornis* (wingspread: 5 meters), reconstructed on the basis of a discovery of its bones (left, below). A present-day crow inserted to show comparative size. (Source: Alan Feduccia)

bird") lived, like the members of the *Teratornis,* in the western part of North America, but in an earlier period, the Miocene (25–5 million years ago). Similar although not as big sawtooths have been found in England and New Zealand; they were thus spread over the whole world and may have looked like the albatrosses of our day, and lived like them. But their closest relatives are the pelicans, cormorants, frigate birds, rails, and cranes, all of which belong to the order Pelecaniformes or webfooted birds.

Tension is at its height: *Teratornis* and the sawtooth battle for the world's record. Which one is going to win, which one is the biggest?

For a full decade it seemed to be a dead heat. But then came a discovery on the east coast of North America, in Maryland, of another sawtooth: *Pelagornis miocænus,* the "Miocene bird of the sea." This species had been discovered on the other side of the Atlantic—in France—as early as 1857, the paleontologist M. Lartet having published a description of a gigantic upper part of a wing bone. Unfortunately, the bone had been badly damaged in excavation (which occurred without Lartet's participation), and between the pieces that remained there were no connecting parts. Lartet had to be satisfied with provisional placements and indicated a possible minimum length of 58 centimeters. According to Lartet the same bone in the modern albatross *Diomedea exulans* measured 41 centimeters. But Lartet did not estimate the width of the wing in his discovery, and the gigantic French bird was forgotten until once again it became the topic of the day when there was talk of the new find (which at the time of writing has not yet been described in the literature; on the other hand, you can admire a half-size model of the bird in the Calvert Marine Museum in Maryland).

The new find is a well-preserved skeleton that shows,

according to the paleornithologist David Bohaska, that *Pelagornis'* wing spread was greater than *Osteodontoris'* and was presumably around 6 meters. Its "upper armbone," so to speak, must have been longer than Lartet's estimate by 20–30 centimeters or more.

The contest has been decided; the giant Atlantic bird is the world champion, and the teratorns will have to concede without further struggle.

Its triumph was short-lived. In 1980 a team consisting of Kenneth E. Campbell from California and Eduardo P. Tonni from La Plata announced the discovery of *Argen-*

Argentavis magnificens pictured in full flight, compared with a sea eagle, a crow, and a wren.

tavis magnificens—Argentina's magnificent bird—which was a *Teratornis* with a wing spread of over 7 meters. In other words, it was twice the size of the first known *Teratornis,* the one once regarded as an unrivaled flying giant. Next to *Argentavis* it would seem like a crow beside an eagle.

Argentavis is only a skeleton, incomplete but still so well preserved that the viewer can get a good idea of the bird's size and shape. Its shape is much like *Teratornis'*; a fully preserved wing bone makes it possible to estimate the span between the wing tips at 7 to 7.6 meters (about 23–25 feet). The skeleton was discovered in layers dating from the latter part of the Miocene period (its age is roughly estimated at 10 million years) in the province of La Pampa in Argentina. So *Argentavis* lived at an earlier time than the North American *Teratornis* birds, which flourished between 3 million and 10,000 years ago.

Argentavis was twice the size of *Teratornis merriami,* the largest flying bird known before 1950. The latter's wingspan of 3.5 meters was only half of *Argentavis'* 7–7.6 meters. This implies that *Argentavis* had about four times the wing area of the *Teratornis* and that it weighed eight times as much. The experts will no doubt arrive at a weight between 80 and 120 kilos (176–240 lbs.) for *Argentavis,* which puts it in the same weight class as an ostrich. When standing on the ground its eyes would be on the same level with a fullgrown man. Nevertheless, there is no doubt that it could fly; the powerful wings were formed just like those of other airborne birds. In one of the wing bones, the one corresponding to our elbow bone, you can even see the places where the wing's contour feathers were attached. (These are feathers that help mold the wing into an aerodynamic shape.) The wing pinions must have been enormous—their length has been estimated at 1.5 meters and their width at 16–17 centimeters. With the thickness of the shafts about 2–3 centimeters, this would make a quill pen of heroic proportions.

Campbell thinks it possible that *Argentavis* flew in the same way as our modern condor, for the most part gliding through the air on motionless wings, with brief moments of flapping flight. He thinks that an *Argentavis* seen from a distance would be strongly reminiscent of a giant condor, while closer up it would make you think of a long-beaked eagle. And this observation provides a key to the teratorns' way of life.

For common to them all is the beak, which is different in shape from the vultures'. The beak in the vultures (including the condor) is short and powerful. The head and neck are completely featherless. Thus the vulture can stick its head into a carcass and tear loose gobbets of meat without soiling its plumage. In teratorns the beak is longer, slimmer, and designed for much more versatile use, presumably as the result of an adaption from carrion-eating to predation. The prey, most probably a small mammal, was caught in the long beak, manipulated into a convenient position, and then swallowed whole. In harmony with this theory is the fact that the *Teratornis* birds did not have the small feet of the vultures but very powerful ones, adapted for use on the ground. On the other hand, their feet were not designed for gripping or grasping like those you find on raptors that seize their prizes with their talons. So it is thought that the teratorns hunted mainly on the ground and used their wings to transport the quarry from the hunting area to the nest.

The teratorns thus should not be represented with an exterior like a vulture's. They probably had a feathery head, looking something like an eagle. Still, no one believes that they were related to the diurnal raptors such as eagles, falcons, buzzards, Old World vultures, etc.; anatomical details point rather to kinship with the storklike birds, which include not only storks but also herons, ibises, and New World vultures.

Campbell and Tonni have described in some detail the environment in which *Argentavis* lived. Grassy fields seem to have been prevalent, to judge from the teeth of the mammals whose fossils were found in the same layers as *Argentavis*. Leaf-eaters are almost completely absent, although they were common in the preceding time period. Many fossil remains of ants of the genus *Atta* (leaf-cutting ants) also suggest that a warm, dry climate prevailed at that time. These ants were specifically the kind that lived in savannas with an extensive dry season. So the climate was evidently subtropical and probably very windy, which must have suited these giant birds well; like condors, they may have become airborne simply by opening their wings to the wind.

The skull of *Argentavis,* over 55 centimeters long, had jaw joints constructed in such a way that it was possible to swallow prey more than 15 centimeters in diameter. Among the fossil mammals whose remains were found in the same layer, the predominant species is called *Paedotherium borrelloi*—the remains from this animal make up no less than 64 percent of the whole. It was a small rodentlike ungulate, or hoofed animal, that belonged to a now extinct South American order of ungulates (Notoungulata). It was about as big as a hare and may have been the ideal prey for *Argentavis*.

With *Argentavis* we find that a new kind of evolution within the bird species reached its culmination. It is one example of beautiful and seemingly perfect beings that we encounter in the distant past to show us that the products of yesterday's evolution need not be inferior to those of today. When you consider the technical flight problems

facing the organism because of such a tremendous increase in size, you realize that in some ways development had gone further than in any of the modern avians. Thus we could claim that the teratorns (like the osteodontorns) had reached a more advanced—or higher, if you will—stage of evolution than any bird of the present day; they had become supremely adapted to overcoming the obstacle of gravity. Still, they died out; perfection in any field is no guarantee of survival.

The seagulls exemplify the opposite contention. They are fair at almost everything and in no way perfect: they are all-around birds. They are good fliers but a falcon is faster, a crow turns faster in the air, an albatross is more skilled at soaring. The gulls are the birds' decathlon champions, which is why they are so successful.

But we must not get the idea that *Argentavis* became extinct just because it was "too big." Like other extinct species, large and small, it died out because changes in the environment robbed it of its basis for existence. In just which way we do not know—there are hundreds of alternatives. The first to come to mind, just as with the sabertooths, is that their favorite prey died out.

Although being in the presence of this giant bird is mind-boggling, I still think Campbell is a trifle imprudent to write that *Argentavis* was "the biggest bird that has ever flown." I can almost hear the echo of a belly laugh from *Teratornis merriami, Teratornis incredibilis, Osteodontornis orri, Pelagornis miocaenus,* and perhaps also some unknown monster biding its time in a layer from the Tertiary period.

CHAPTER ELEVEN

Eve in Africa

We carry our inheritance with us in every cell of our bodies. Enclosed within a membrane, it is located in the heart of the cell, its nucleus, in the form of chromosomes: long threads of deoxyribonucleic acid, DNA. The chromosomes' command codes determined how we were to develop from a fertilized egg to a grown individual and how we are to function as biological specimens. When the sex cells are formed—the egg in the woman, the sperm in the man—the combined substance is divided in such a way that the new individual will receive one-half of his or her chromosomes (and thus one-half of the DNA) from the mother and one-half from the father. So what the child inherits from mother and from father are of equal weight —at least in the chromosomes.

But the cell also contains DNA that is not stored in the chromosomes in the cell's nucleus but is found in the cytoplasm, the "nuclear sap" that surrounds the nucleus. It is found, for instance, in those parts of the cell called mitochondria. They have a special history. About a billion and a half years ago the first cells with genuine nuclei appeared, that is, cells with DNA collected in a cellular

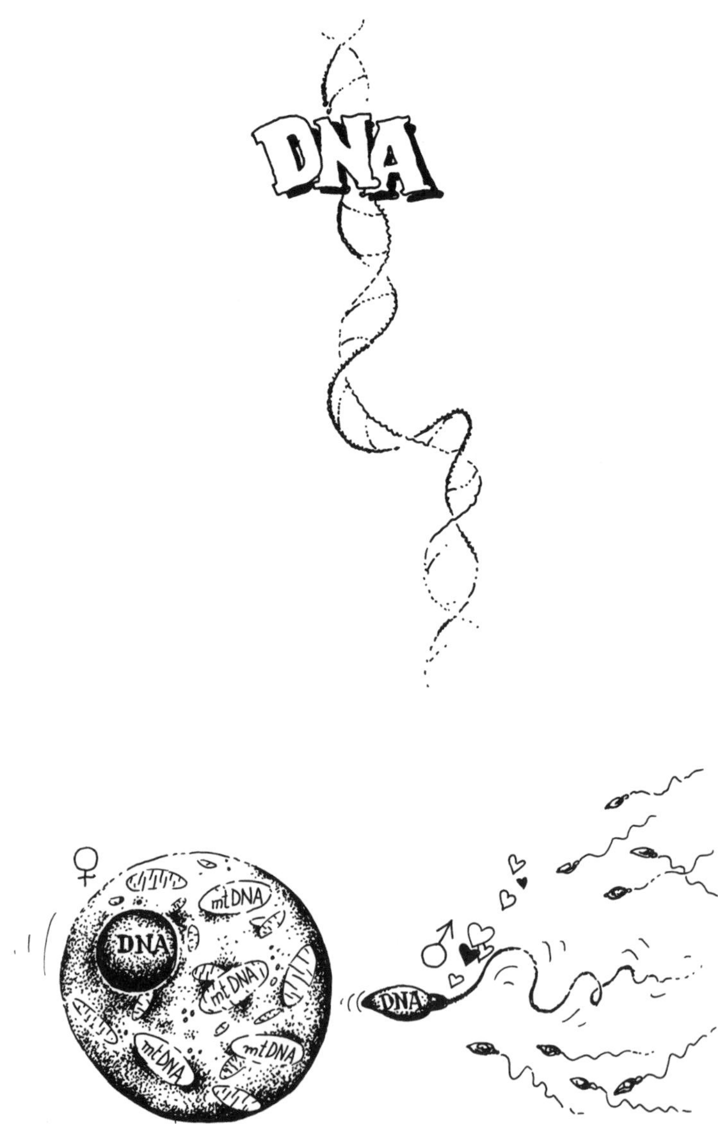
DNA
DNA
mtDNA
mtDNA
mtDNA
mtDNA
DNA

nucleus. Such cells are called eucaryotes, in contrast to the procaryotes (bacteria and blue-green algae), among which the DNA is located in the cytoplasm. At the time of the first eucaryotes there were also, among others, bacteria of a type that "respired," which means that they used an acid to burn food and produce energy. That is the kind of bacteria that invaded the eucaryote cell and now lives in an intimate and symbiotic relationship with it. The mitochondria still conduct most of the chemical energy production of the cell. (There are also eucaryotes that lack any mitochondria, but they are comparatively rare.)

The bacterium, of course, carried along with it its DNA, and it is this that now is known as mitochondria-DNA, or mtDNA. This DNA too is inherited from one generation to the next—but only in the female line. The sperm of the man are tiny; they actually consist of only a nucleus and a long tail, with the aid of which they swim over to the egg that is to be fertilized. Their mitochondria are aborted along with the tail and do not enter the egg. On the other hand, the egg, which is lying still and doesn't need to move, is a complete cell, a giant in comparison with the sperm, and contains mitochondria as well as other cellular parts. That is how it happens that sons as well as daughters inherit all their mtDNA from the mother and nothing from the father. One might say that mtDNA continues to reproduce itself asexually, just like the original bacterium from which it is descended.

The mass of mtDNA is small, a fraction of the mass of the chromosomes. As inherited matter it plays a modest role. But to those who conduct research in the field of evolution it is of great interest, for two reasons.

First, mutations tend to accumulate much more quickly in mtDNA than in the DNA chromosome; on the clock of evolution, mtDNA is the second hand while the DNA

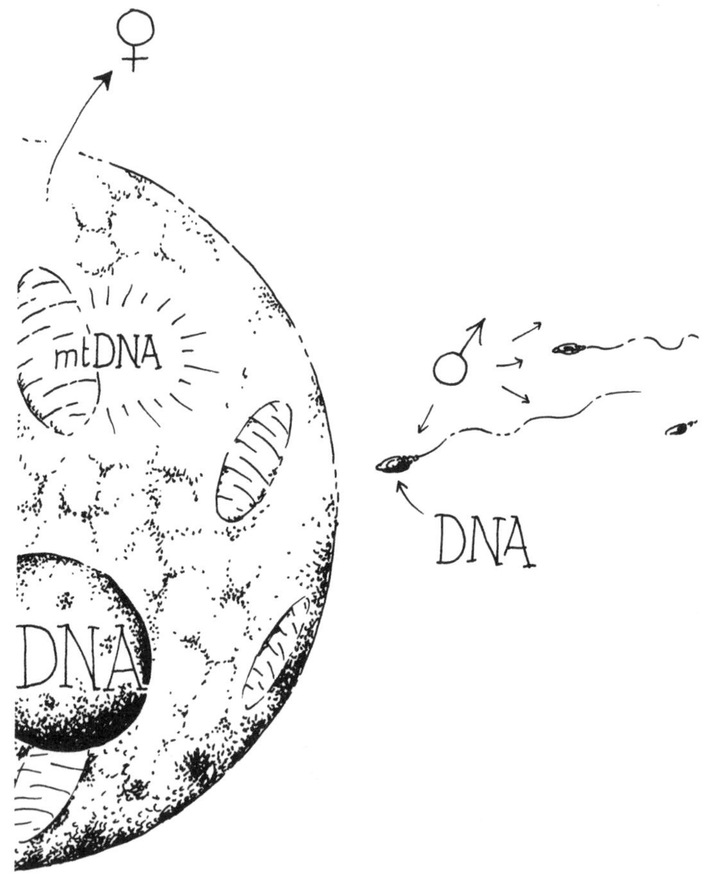

chromosome is the hour hand. In a way, that makes it possible for us to see evolution as if under a microscope; details that were not visible before are greatly enlarged. Where the DNA chromosome needs millions of years to change noticeably, the mtDNA is ticking away with thousands of years as its measurement of time.

The other reason is that what's inherited passes down only in the female line. The chromosomal genes, which are inherited from both parents, are mixed in every new generation; the variation thus brought about is working material of natural selection. But this implies that it is not possible to trace the history of a given DNA segment by itself, since a recombination takes place in every new generation. For mtDNA, however, it is possible. Consider: we have four grandparents, but from only one of them, our mother's mother, do we obtain our mtDNA. Among the eight forebears in the previous generation there is again only one, the mother of our mother's mother, who has left her mtDNA for us to inherit; and back another generation, her mother again is the only one of sixteen ancestors who qualifies on this score—and so on. But while our entire mtDNA is derived from our grandmother's grandmother, all in the female line, we have inherited from her a mere sixteenth of our DNA chromosomes.

So mtDNA is inherited down the line unchanged, identical from generation to generation. We all have an mtDNA like that of our mother, maternal grandmother, maternal great-grandmother, etc.—unless a mutation occurs. When that happens, the mtDNA continues to reproduce in its new, mutant form.

From all this (plus, of course, intensive scientific work) it is now clear that the ancestress of humanity lived and bore children in Africa about 200,000 years ago. That is the result arrived at by a research team from the University of California at Berkeley. The team consisted of Rebecca L. Cann, Mark Stoneking, and Allan C. Wilson; their report appeared fairly recently in the journal *Nature.* They investigated the mtDNA of 147 individuals, representing various geographical regions: Africa, East Asia, South Asia, Europe, Australia, and New Guinea. The 147

mtDNA
mtDNA
mtDNA
mtDNA

individuals represented 133 different types of mtDNA; the individual differences varied between zero and 1.3 percent. (In this case 1 percent implies that of one hundred alleles that connect the two DNA chains, one allele will have been replaced by a new one, through mutation. In fact, of course, DNA contains millions of alleles.)

The degree of relationship between different individuals is made clear by the similarity in DNA. To simplify, suppose that we have four individuals with the following gene arrangement:

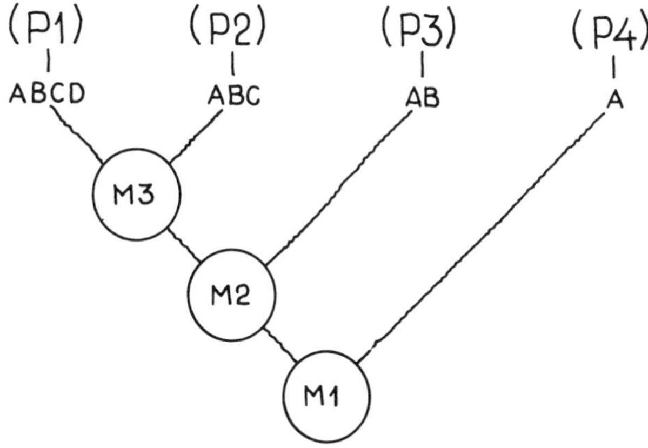

They must all be descended from a common ancestress, whom we shall call M1, with the gene A. For P1, P2, and P3 we must assume in addition a later common ancestress, M2, in whom the genes A and B were both found. B had been created through mutation. At a still later stage lived a common ancestress M3 for both P1 and P2; in her the gene C appeared as the result of a mutation. From her came the genes A, B, and C to P1 and P2. So the specifications outlined here result in an evolutionary diagram like the one above.

But mark well—it did not necessarily happen this way. This is merely the most economical interpretation, the one that requires the smallest number of mutations. Our analysis here is carried out in accord with Ockham's Razor, also known as the parsimony principle.

This example also shows that variations in mtDNA tend to increase as time goes on. In other words: the older the population in a given area, the greater is the variation within the mtDNA. The greatest variation is found in Africa south of the Sahara, and thus quite probably that was where the real primordial mother lived.

How long ago? We know that human beings have lived in New Guinea for at least 30,000 years, and in Australia for at least 40,000 years. On the basis of these numbers we can estimate the speed with which mutations tend to accumulate in mtDNA. The result is between 3 and 4 percent of DNA for each million years. We can then estimate the age of the primordial mother at between 140,000 years (at 4 percent) and 290,000 years (at 3 percent); so the California group accepts 200,000 years as a fairly close estimate. Of course, we can only settle on an approximate magnitude, not decide on a definite age.

You often find that certain regions seem to have been colonized by two or more "types" of people. That doesn't necessarily mean two different waves of settlement, although it may. It may also be because the women who belonged to the original settlement were distantly related to one another in the female line. We must never forget the special way that mtDNA is being transmitted!

The results of the Berkeley scientists' research are of

great interest with regard to the evolution of modern human beings. They seem to decide conclusively between two disputed claims, two opposing theories. One theory is the "polycentric" one, according to which modern humans evolved, so to speak, on a broad front, from local ancestors in Africa, Europe, and Asia. The other theory is the "monocentric" one, according to which there was a geographically limited primordial home, where modern humankind evolved and from which it emigrated and colonized the whole inhabitable world, displacing the older human types (as for instance the Neandertal people in Europe and Peking Man in China.) The polycentric theory was dominant in the 1960s and 1970s; probably it still has some adherents, although the work of Cann and her colleagues ought to have been its deathblow. The monocentric theory is the only one that is now scientifically respectable, which, of course, is pleasing to those of us who have always, on biological grounds, found it to be more probable.

Was Adam a contemporary of Eve's, and was he too an African? We could study this problem by investigating the Y chromosome. This chromosome is the male sex chromosome—the female is the X chromosome. A woman has two X chromosomes, while a man has one X and one Y chromosome. This implies that the Y chromosome will always be transmitted from father to son. It ought to be possible to conduct a study of the male line like the one that was made of mtDNA in the female line. It is hardly probable that Eve and Adam in this sense will prove to be contemporaries; it is more probable that they lived in the same continent.

I mentioned that the invading peoples of the modern type—we call them *Homo sapiens,* our own species—pressed in on and displaced the older peoples whom they encountered in, for example, Europe and China. How did that happen? Did it come about, as has been assumed, through crossbreeding of the races between those who were the old occupiers of a territory and the newcomers? In that case, you would think that the crossbreeding in many cases would take place between native women and immigrant men. (That is the usual pattern.) Then we ought to find in the European and Asiatic mtDNA investigations greatly disparate gene sequences that cannot have been derived from the common African primordial mother. We must remember that mtDNA is not passed on like an infection but is transmitted from person to person quite independently of other mtDNA. The Berkeley group found nothing like that. That suggests that crossbreeding did not take place (alternatively, it might mean that crossbreeding did occur but that the progeny were not fertile or that the hybrids were so poorly adjusted that little by little they died out without leaving offspring). What is important, however, is the fact that the Neandertal people bequeathed nothing to the gene pool of present-day Europeans, no more than Peking Man did for the Chinese of our time.

Was "Eve" the first *Homo sapiens?* There is no special reason for us to believe that. All we can say is that she belonged to a human population, containing many other women who may also have been our ancient mothers, but in that case in some part of the male line. In any event, they must surely have made contributions to our inheritance in the form of chromosomal DNA. Nevertheless, "Eve" is the last one in the female line who is the primordial mother of all people now living. But that is not the same as saying that she was the first *Homo sapiens.* She may have belonged to our own species, but she—or rather the population of which she was a part—may also have belonged to the species that was the primitive form of *Homo sapiens,* that is, *Homo erectus.* On the other hand, it is quite possible that the transition from *erectus* to *sapiens* occurred much further back in time than "Eve," so that our species had existed for some time before this primordial mother lived. On this mtDNA can give us no information; it offers us a genealogy but no classification.

Instead we can continue studying the "hardware," the historical records represented by fossil bones. They accord well with the theory of an ancient African population. True, the African chronology, so far as we can determine it now, is not very exact, but several *Homo sapiens* remains have been found that are thought to be 100,000 years old or more, and there have been even older finds that may have been transitional forms between *erectus* and *sapiens.* No such transitional forms have been found in Asia and Europe. In Asia it appears as if the local *erectus* disappeared and *sapiens* arrived in its stead, and in Europe the course of events is even clearer: Neandertal dies out and is succeeded by the modern *sapiens.* It is not impossible that the last groups of Neandertalers in Europe carried *sapiens* genes—certain peculiarities in their skulls may be indications of that—but it doesn't seem as if genes were transmitted in the opposite direction, to judge from the mtDNA.

So it is most probable that Africa was our primordial home—that is, of our species *(Homo sapiens).* Even in another sense Africa is our primordial home. As far as is known, it was here that the first members of genus *Homo* came into existence 2 million years ago. And *Homo*'s precursor, the genus known as *Australopithecus,* also lived in Africa; the oldest known so far have been found in layers almost 4 million years old. Our closest relative among the apes, the chimpanzee, is also an African species. If we go even further back in time, to the first anthropoid apes—

they lived almost 35 million years ago—once again we find that the first ones lived in Africa and that they migrated to Europe and Asia about 17 million years ago. History has repeated itself many times.

It is interesting to see that even other kinds of animals have specific primordial homes, where they evolved and then spread out to the other continents. The original home of the horse family was North America; time and time again new and more highly evolved variations on the horse theme found their way to the Old World, and also, little by little, to South America. The same is true for the camels. So our own history is one example in a grand design of infinite variety changing always in cadence with the long evolution of life.

CHAPTER TWELVE

Dog Cat and Goat Antelope

I could hear the irritation in her voice when the head waitress at the Tunis Hilton lost her dignified mien and said laughing, "You like *brick* too much, sir." Then I realized that for four days in a row I had ordered the same dish, *brick à l'oeuf,* partly because I liked it but also because I could eat it quickly and return to the analysis I was doing. During the day I was occupied with the remains of 12-million-year-old antelopes on the premises of the Geological Survey of Tunisia back in the city, and in the evenings I was analyzing my data. The antelopes numbered 656, all belonging to the extinct species *Pachytragus solignaci* (Solignac's fat goat), and we had collected them all in three seasons of excavation.

Of course, it was not a question of 656 whole antelopes. Many were represented by jaws or even only teeth. Still, it was what one might call a respectable statistical sample of Solignac's little goat antelope, about one meter high and named in honor of a French geologist. The interesting

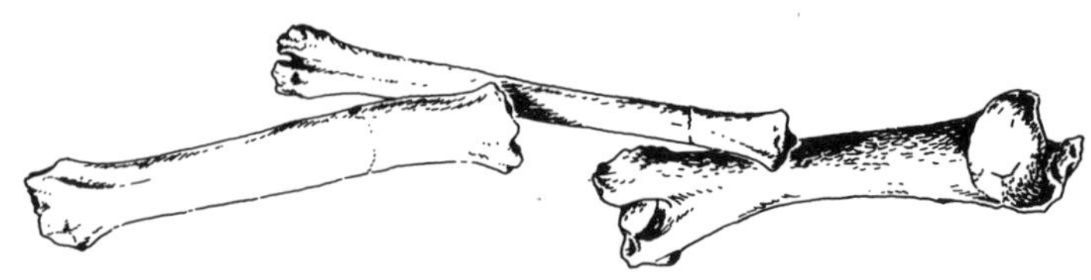

Bones of the goat antelope

thing was that all the animals had died at the same time of the year. That was especially clear for the calves. The youngest ones, with healthy milk teeth and the first permanent molar all in place, had died at an age a little under one year. In the second group, which was about a year older, the milk teeth were completely worn or had fallen out, and two molars were in place. Among the three-year-olds all three molars were in place and in use. (There are never more than three molars in each jaw.) Since there were no intermediate stages—all the dead ones were uniformly one, two, or three years old—they must all have died at the same time of the year, probably during a dry season.

That must also mean that the older specimens—in whom the adult teeth were fully grown, so that determining their age on that basis was impossible—form similar age groups with one-year intervals. But it ought to be possible to determine their ages from wear and tear on the teeth, and that was what I was engaged in.

The result of the analysis was that it was indeed possible to determine the age at death of every one of those 656 antelopes that lived in North Africa 12 million years ago. And we could put together an actuarial table for the antelope population—in about the same way that the life insurance companies do it for us—and calculate the mortality rate for each age (overall it is about one in four, 25 percent, while human beings, in their prime, have a mortality rate of a few percent). In addition, we could determine their life expectancy (the years they have yet to live), which sounds rather paradoxical for animals that have been absent from the surface of the earth for millions of years. A one-year-old *Pachytragus solignaci* could expect a remaining lifespan of three years and five months. So these shadowy beings from the distant past emerge and seem to whisper, "We lived too once upon a time," since their pattern of growth and spread over a given territory was the same as that of their modern relatives. The mortality rate increases with age; it grows to 40, 60, 80, and finally 100 percent for the old ones, which after ten years have worn their teeth down to senile stumps.

Which ones die young, which ones live to an advanced age? Here we can observe natural selection at work. In *Pachytragus solignaci,* as in other species, the teeth vary in size among individuals—but the variation grows smaller the older they become. That is because the extreme variants—unusually large or unusually small—tend to die earlier than those that are closer to the norm. The average goat antelope lived the longest. That it had something to do with the teeth's function is proved by the fact that selection comes into play as soon as tooth is involved in the chewing process but not before, that is, not so long as the tooth is being formed in the jaw and then erupts.

When the selection favors the "normal type" and tends to discard the extreme variants, we speak of *stabilizing selection.* That indicates that the species is well adjusted to its environment. In other conditions we encounter *directed selection,* which favors variants that diverge from the normal type—for instance (when teeth are involved), larger,

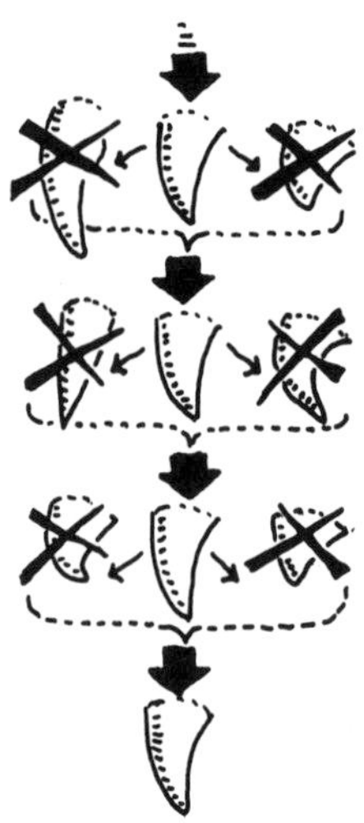

Stabilizing selection (above)

smaller, broader, narrower, etc. Selection of this kind has been noted in the Ice Age, cave bears, among others.

Of course, I had no time to dwell on all this during the days in 1970 in which I dined on the finest products of Tunisian cuisine. The work on *Pachytragus* was finally published in 1983, in the journal *Paleobiology*.

It all started in 1966, when a good friend and colleague, the paleontologist Peter Robinson from the University of Colorado, asked me if I had any desire to take part in an expedition to Tunisia. It was to be underwritten by the Smithsonian Institution. Peter had approached me because of certain vague reports in the paleobiological literature to the effect that some French scholars, quite some time ago,

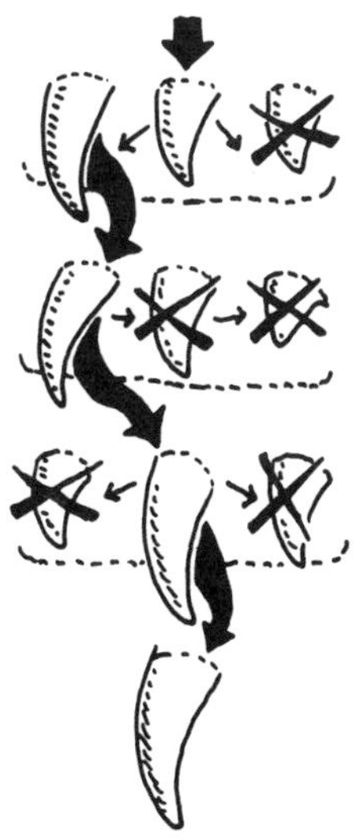

Directed selection (below)

had found mammal remains from the Miocene period (25 to 5 million years ago) in one part of Tunisia. He knew that I was interested, since my Ph.D. thesis had been on population dynamics among Chinese mammals in that period. Of course, I agreed immediately.

I arrived in Tunisia in February 1967 and found that the expedition consisted of only two men, to wit Peter and I, and Peter had been delayed for a few days. Broke and devoid of compunctions, I installed myself per instruction in the Tunis Hilton, and felt like some kind of international con artist, until Peter showed up. Our first task was to note the locations of the most productive discoveries. And so we entered into a most pleasant relationship with the Geological Survey of Tunisia, which helped us with maps and laboratory equipment, and whose museum in time would receive the materials we collected.

All we had to go on was a rather old geological map of Tunisia and reports of a number of discoveries of Miocene mammals, partly in northern Tunisia and partly in the vicinity of Tozeur in the south. Our prospecting in the north was unencouraging, and we set course for Tozeur. My hopes regarding this town and the giant salt flats called Chott Djerid were not disappointed, although we found no fossils aside from the ever-present oyster shells, originally a part of the chalk layer but now crumbled into dust.

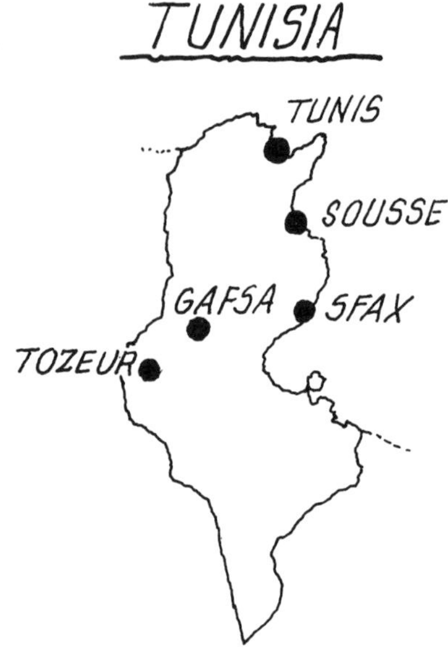

Perhaps the reader is not acquainted with the procedures involved in discovering a place where fossils may be found. The work consists mainly of walking and looking at the ground. If you're lucky, you might happen to notice a fragment of a small bone that has come from one of the chalk deposits in the matrix. The first bone may be a clue to the whereabouts of a rich mass of fossils. But you might also have to walk a long time before encountering anything of interest. We spent three weeks at such prospecting in 1967, covering from 10 to 20 kilometers a day, in very rough terrain. It helps keep you in good shape.

The main difficulty is in noticing that little bone fragment in the first place. It is hard enough when you're walking on your own two feet, but according to an account by Abbot René Lavocat it is much more difficult from the back of a camel. The camel's motion makes you rock and sway, back and forth, which makes the ground seem to fly out from beneath you, first in one direction and then in another. If you actually manage to notice a fossil, the ground is usually so filled with stones that the camel refuses to lie down so that the rider can dismount. To retrieve the fossil on one occasion Lavocat would just drop a hammer to mark the spot. He could then guide his camel to a suitable place to dismount and then walk back to the hammer. Unfortunately, he said, the outcome would often be that he could find neither the fossil nor his hammer.

Abbot Lavocat's account, delivered in a booming voice with sweeping gestures, was listened to with a lively interest by, among others, Professor Bryan Patterson of Harvard University. He remarked that Lavocat ought to be on TV. "Yes," replied Lavocat, "and Professor Patterson will play the role of the camel!"

We learned a whole lot about the geology of the area before we had identified our most promising locations.

Capsian flints

Toward the end we became so skilled that we could identify Miocene layers at a distance of several kilometers. Our most important field of operations, however, became a large basin called Bled Douara, situated among the mountains west of the town of Gafsa, which the Romans called Capsa. In this basin fossil-bearing strata from the later Miocene are found. Stone Age people lived here at a much later date; the ground is covered with flint objects that were clearly worked or fashioned by human beings: long, elegantly formed chips, from the late Ice Age culture known as the Capsian.

Northern Tunisia has some fairly high mountains, the last offshoots of the Atlas range. They belong to the "Alpine" generation of mountain, formed during the middle and late Tertiary times. In the southern part of the country the upheavals in the earth's crust have been minor and have led only to an undulating topography with low, undramatic rises. Below the slopes of the rises, in the so-called anticlines, the most recent strata have been worn down and reveal marine layers from the Cretaceous and Jurassic periods. Sediments from the Tertiary period are found in the flanks of such rises in the terrain. The oldest of these sediments are of great economic importance since they contain rich deposits of phosphates. Still further in toward the sloping basin, in the synclines, are found the Miocene strata, located in its fairly steep sides. The folds in the mountain were created in the late Miocene, and the resulting synclines have since been filled with younger layers, all in a horizontal position.

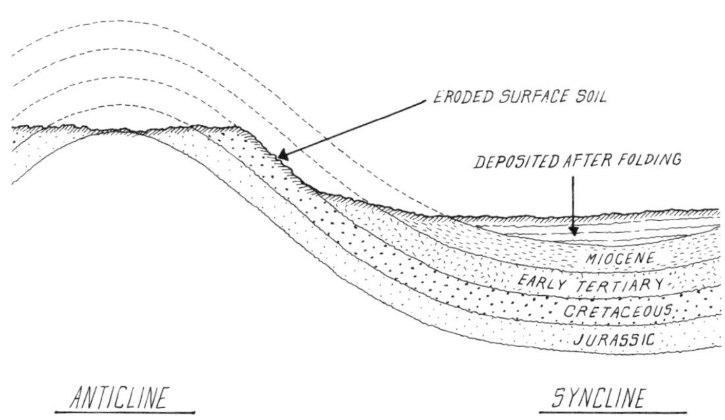

The Bled Douara basin is such a syncline. Along the southern edge of the basin you find Tertiary layers, horizontal when they were created but now steeply descending toward the center of the basin. Thus, when you proceed from the southern mountains northward toward the basin you will cross progressively younger layers. The Miocene layers consist mainly of loose sand, which in certain sharply restricted areas contain large numbers of fossil bones. That is where we worked during the 1968 to 1970 seasons, and the expedition grew in size with the addition of a few participants from the United States, Spain, and Finland. We also hired a work force of bedouins living in the area, close to 50 people in all. That made us popular, since unemployment is a scourge even in the Tunisian desert.

Our activities were watched with great interest by our closest neighbors, a bedouin family, whose 12-year-old daughter, Anis, proved to be fantastically clever at discovering new fossils throughout the area. The girl used to come to us with a secretive expression on her face and motion to Peter or me to follow her, whereupon she would lead us to some point in the terrain, stick her hand in the ground, and lift out a fossil jaw with the elegance of a magician pulling a rabbit out of a hat. And it was far from faked: when we started to dig we would always find a sizable collection of bones. One of the places Anis detected was to produce the greater part of our collection of goat antelope bones. It is with great justice that one fossil animal species bears the name of Anis.

The expedition had no lack of merry interludes; a typical one may be found in my diary for March 1968: "Have just taken a bath to get rid of all the sand in my hair, in my ears, etc. We were quite a sight this morning when we trotted out to the latest place we had found fossils with

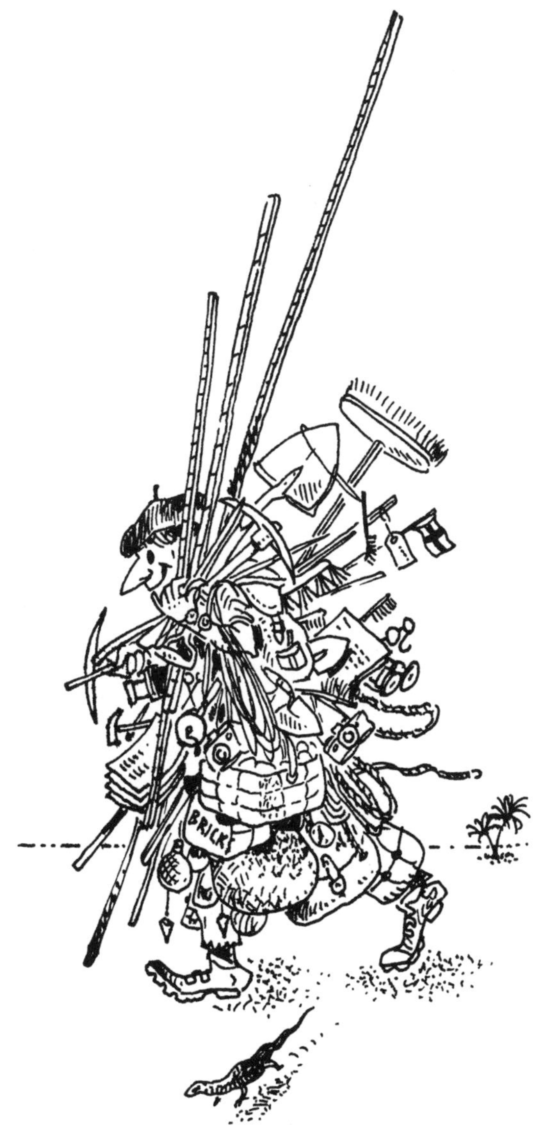
BRICK

all our equipment, pickaxes and spades, as well as a table, tape measure and measuring rod, brushes, trowels, cameras, field bags, and lunch basket (which I was carrying), and in the middle of it all, our Arabian assistant, Abdel Aziz, rushing ahead with our newly purchased wheelbarrow; he was making sounds like a locomotive the whole way, whistling and roaring and saying choo-choo."

But we, and even more so our bedouin friends, were to face one of the recurring human tragedies, an epidemic that broke out in 1969 and ravaged the underfed population. It was up to us to act as ambulance drivers in our cars, which came in handy. Everything turned out well and we were glad that we could help those who had helped us so much.

Since the sediment consists of sand, still of a loose consistency, it was easy to work with; the fossil bones were well preserved on the whole but often a bit splintered. We never found any whole skeletons. The remains were probably from animals that had died during the dry season and had been eaten to the bone by scavengers. The rainy season would have arrived with torrents of water that wrenched the individual bones away from any they were connected with, mixed them all into a jumble, and buried them in the sand. At certain levels there are lumps of limonite, bog ore, which can easily be mistaken for bone. Usually, the error is not discovered until you wipe the sand off the object.

The layers may incline as sharply as 45 degrees, which has some advantages: you can get from one layer to the next across level ground. But it has the disadvantage that a rich fossil-bearing layer will rapidly plunge into the earth, and follow-up requires extensive digging. Most of our hired laborers were engaged in this heavy work. Peter, however, was not satisfied with the result and was dreaming of renting a bulldozer, devising many a scheme to that end. He finally succeeded in the 1969 season. The rented bulldozer was transported to Bled Douara on a huge truck and arrived proudly clattering under its own power the last part of the journey.

The bulldozer was set to work immediately and without too much trouble flattened out a hillock that in Peter's opinion was in our way, after which we let it rest for the night. Unfortunately, the driver had parked it in a hollow, and the next morning it refused to move. If it had been standing on a slope it would have been easy to get it moving, but under the circumstances we were just standing there with a dead bulldozer on our hands.

The only thing that can get a dead bulldozer started is another bulldozer. Peter had been trying for over a year to get hold of the one we had, and now he needed another one! Luckily, we knew that a group of Bulgarian experts was working in the phosphate deposits at the far end of Bled Douara, about 25 kilometers from us, and there was a rumor to the effect that they had a bulldozer with them. We paid the Bulgarians a visit, and they were happy to help us out.

At the leisurely crawl characteristic of bulldozers the one that came to our rescue spent, I believe, over two hours to cover the 25 kilometers to our camp, and finally the two monsters were standing face to face. It took only one small shove. With a tremendous roar our machine woke to new life. We cheered our Bulgarian friends, who laughed and waved as they set out on the long return trip after a job that took only two seconds.

After that, our bulldozer was always parked on a slope. But we didn't have as much use for it as Peter had hoped.

In spite of everything, however, our collection did grow at a satisfactory speed. Most impressed were the representatives of the Tunisian Geological Survey, who had not dreamed what treasures we would be able to give their museum. The only thing that annoyed Peter was that we found no remains of small mammals—rodents and insectivores—which were his special interest. Clearly, under the conditions in which the sediments had been created, very small animals, whether mammals or insects, had simply been washed away by the strong currents. My colleague Ann Forsten, on the other hand, was delighted with the great quantity of bones and teeth from the three-toed horse *Hipparion* that we unearthed, since it was the subject of her doctoral thesis. As for myself, the sporadically emerging remains of predators were sufficient to keep me in a state of continual happy anticipation.

The findings of the University of Colorado's Tunisian expedition have since been published by several researchers, mainly in the bulletin of the Tunisian Geological Survey. They include a description of the geology of Bled Douara as well as the places where discoveries were made; in addition, a series of monographs offers systematic description of the fauna. Besides mammals, the fauna include fishes and reptiles—crocodiles and tortoises—and represent many time periods, so that changes in the fauna can be followed over time. The most important change was the sudden disappearance of the three-toed horse *Hipparion* in about the middle of our time sequence. Before this period *Hipparion* appeared in huge numbers, but during the period itself there is no trace of it. We know that *Hipparion* evolved in North America and that it migrated to the Old World (across the then nonexistent Bering Strait) about 12 million years ago, after which it spread rapidly across Asia, Europe, and Africa. The first species to arrive is known as *Hipparion primigenium* (appropriately, "the firstborn hipparion"), and Ann Forsten has

shown that this same species lived in Bled Douara. Thus, we arrived at a date for one of the Miocene sequences at Bled Douara and were able to conclude that our fossils are about 12 million years old.

A few of the remains testify to the presence of mastodons, the huge animals that resembled today's elephants. More common was another curious animal, similar to a hippopotamus, that is known as *Merycopotamus* and also had South Asia as its habitat. It is not a genuine hippopotamus but was a member of a closely related family called anthracotheries.

I have already mentioned the goat antelope *Pachytragus solignaci,* which is represented by numerous fossils—but only in those sediments created before *Hipparion* entered the scene. It is regarded as an early relative of our living chamois, goats, and sheep. Its height, measured from the shoulders down, was about one meter, and it had short, robust horns. Its dental structure may indicate that it was not strictly a steppe- or savannah-dweller but rather an animal that belonged to the wide forests.

Giraffe-like animals, swine, and rhinoceroses are also

found there, all of them belonging to now extinct genera. Deer, on the other hand, are among the missing. Among the herbivores there is instead a species of ostrich.

Carnivores, surprisingly, are represented by many species, although only one or two specimens of each have been found. Among the sabertooths we found the *Machairodus,* invariably accompanying *Hipparion* at several locations in Europe and Asia—but in Bled Douara it appears among the older fauna without *Hipparion.* Identifying it was routine, but I cannot say the same of a canine and a molar (the lower carnassial tooth) of a catlike predator, which I discovered only two centimeters apart in the same place where the goat antelope *Pachytragus* was best represented. After a great number of comparisons and much pondering I concluded that this was a completely new species, so far unknown in paleontology, and I named it *Vampyrictis vipera* (vampire cat, viper). I am almost completely certain that does not belong among the genuine cats but is a kind of nimravid (see the chapter "The Innocent Assassins"). *Vampyrictis* was about the same size as a leopard and had narrow, dagger-like fangs and razor-sharp, thin-bladed molars. We know of many sabertooth cat and nimravid species, but no one of them even approximately has these traits! Perhaps other specimens of this intriguing animal are concealed in the ground at Bled Douara.

Another high point for me was the discovery of an upper jaw, or maxilla, and a radius (one of the two bones in the lower arm) of a "dog cat," a catlike dog of titanic proportions—as big as a Kodiak bear. The big moment arrived when I showed my discovery to a colleague, Johannes Hürzeler, in Basel. He had found a similar animal in strata located at Charmoille in Switzerland, and we immediately took steps to compare our finds. The material from Charmoille proved to be a lower jaw, or mandi-

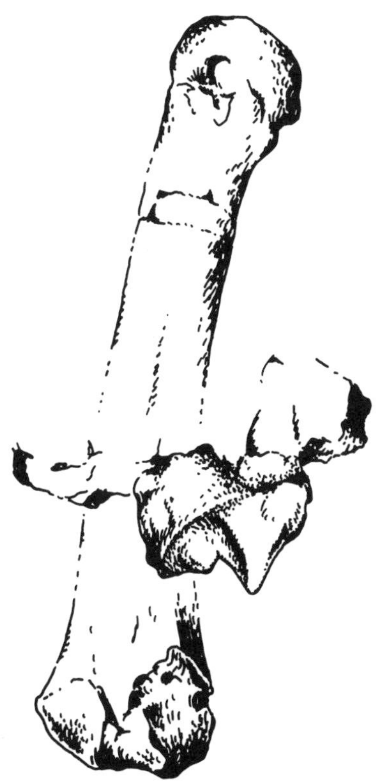

Agnotherium—foreleg and tooth

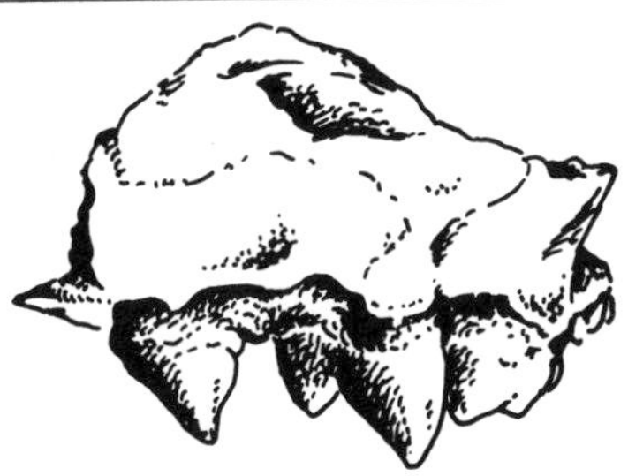

Tunguricitis—teeth

ble, which proved to fit exactly with the upper jaw from Bled Douara, and an elbow bone (ulna) that fit just as perfectly with the Tunisian radius. If it hadn't been completely impossible, we might have thought that the same individual had left half its bones at Charmoille and the other half at Bled Douara. . . . The animal is actually a great rarity; it is the last of a long extinct evolutionary line, and is known as *Agnotherium* ("unknown beast").

Generally the carnivores seem related to Asian animals —*Agnotherium* with its European connection was an exception. An extreme case was a civit of the genus *Tungurictis,* which until then had been known from only a single find in Mongolia; so the second find was made in Bled Douara. We also found unique species of predator, not found in any other place, such as *Vampyrictis.* They may be a native African element among the fauna.

Bled Douara and its fossils represent only a brief moment in the long history of the African fauna. There are thousands of places where mammalian fossils have been found, many of them much more fruitful and better known than Bled Douara, and all have a history filled with the joys of discovery. Together they provide us with a detailed panorama of the development of life. I have chosen to speak of Bled Douara because I was privileged to take part in an investigation that gave us so many memories of a beautiful country, of hospitality, and of good fellowship. I especially remember the trip home after a day's work in the field, a 45-kilometer commute in the evening. Driving toward Gafsa from Bled Douara, when you reach the mountain ridge that marks the eastern end of the valley the view is breathtaking. The wide valley is utterly flat, bathed in the evening sunlight, dark clouds reflected below in a moving pattern of shadows. In the distance you can see the Gafsa oasis spread out about 20 kilometers

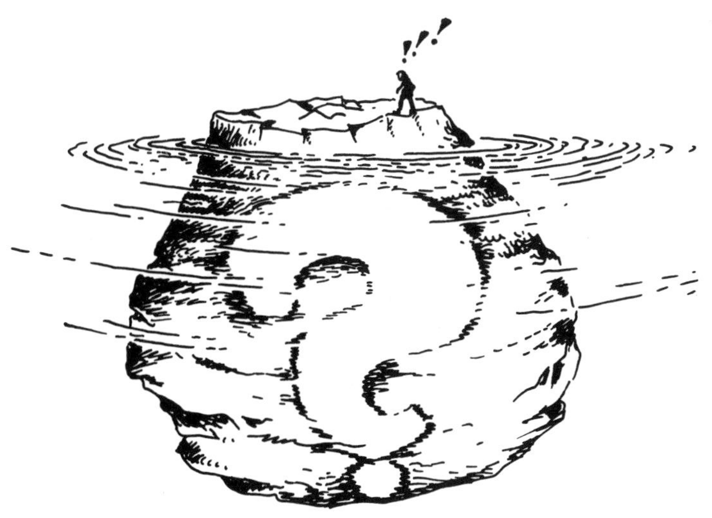

away, with the Djebel Ben Yunes mountain seeming like another shadow in the haze beyond. A little to the left is a smaller oasis; 15 kilometers away from the hotel, it looks tiny below the rust-red slope of Djebel Ben Yunes. In the foreground is a landscape that can only be called "badlands," which you must cross, and to the left the peculiar tiger-striped flanks. They were carved out by a score of small parallel valleys created by erosion, and in the evening light they make a fantastic pattern. Darkness falls, and a half-hour later we find ourselves in another world, washing off the day's dust, getting together for a meal about nine in the evening at the hotel's fine restaurant. It is almost impossible to imagine doing fieldwork under pleasanter circumstances.

What we know is the tip of the iceberg. Infinitely more remains to be discovered. In Algeria the animal life is slightly younger than that of Bled Douara ("slightly," that is, in the geologic sense of the word), although it has quite a different composition. For example, all our Tunisian carnivores are absent from Algeria, but instead there is a giant hyena not found in Tunisia. During all these aeons of time the world has been constantly renewed, is always young and luxuriant beyond anyone's comprehension. The paleontologist is a spy and an observer, a fisher in the stream of life.

CHAPTER THIRTEEN

The Steppe Bison of Beringia

Just 36,000 years ago a steppe bison died in Beringia, the last of its kind. Beringia is the name of the broad tract of land that united Siberia with Alaska during the Ice Age, a hundred-mile-wide bridge that became dry land when the sea level dropped, the water locked in the massive ice-sheets covering the land. The bison died near the site of present-day Fairbanks in the lowlands of central Alaska.

It was a nine-year-old animal, not a particularly large one. It belonged to the now-extinct Ice Age species *Bison priscus* (the name simply means "ancient bison"). The European wisent of the present day is probably a descendant of *Bison priscus,* the main difference being that the horns of the living wisent are considerably larger.

The species was common during the latter half of the Ice Age, one of the hardy steppe animals that ranged over Europe, Siberia, and the ice-free parts of Beringia; its eastern limit was the inland ice on the North American continent. Skeletal remains of steppe bison are often found

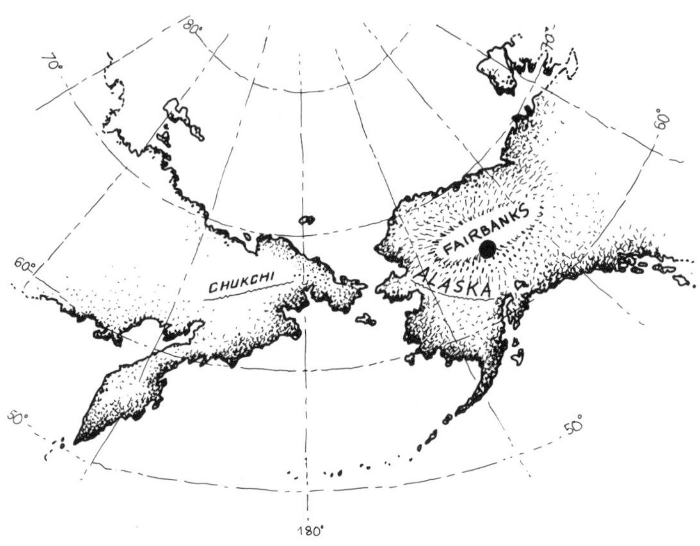

among the cultural litter of the Neandertals and their successor in Europe, Cro-Magnon, so it may be assumed that Ice Age people hunted this species for food. Certainly it was one of the favorite animals of the Ice Age artists, and innumerable representations are known—sculptures, engravings, paintings. The material includes the famous polychromes of Altamira with their magnificent renderings of the steppe wisent.

Let us pinpoint the situation at year 36,000 before present. It falls broadly within the Last Glaciation (which commenced some 100,000 years ago, and ended 10,000 years ago) but, as it happens, not at a time of maximum cold. There is evidence that the climate, although colder than now, was not fully glacial, and that most of the inland ice-sheets had vanished temporarily. In fact there

may have been a gradual warming trend at the time. Some 10,000 years later that trend changed and the last great glacial phase began.

It was the time of the last Neandertals in Europe. Humans of modern type were already in existence elsewhere, and their vanguards were probably on the thresholds of the Neandertal land. About 5,000 years later, the last Neandertals seem to have vanished.

Beringia, at the time, was partly inundated. Still, its American part harbored a remarkable fauna of large mammals, including such Eurasian immigrants as mammoths, steppe bison, moose, and grizzly bear. Not to mention the lion, remains of which have been found in abundance in Alaska. Lions are intimately involved in our story, since a pride of them killed our Alaskan steppe bison.

Cattle, in dying, tend to topple. A bison, on the other hand, usually sinks down on its belly, and so did this one. The lions went to work. With their sharp teeth they cut up the tough hide along the animal's back and stripped it down its flanks, exposing the meat, of which most was eaten. It seems that ravens too were picking at the dead body.

Before long, however, the winter cold put a stop to this. The meat froze and became hard as flint. One lion, hopefully trying to prise off some of it, broke its tooth and left a sliver of it in the meat. And so the half-eaten wisent remained, a frozen, unrecognizable mass, in the same spot where it had been killed.

The bison had died at the bottom of a small gully that had been excavated by a creek when the climate was warmer. Now it was cold and the creek had dried out. But it could still do a job that, many thousands of years later, would fascinate the descendants of the bipedal beings then poised to enter Europe. Mixed in with the gravel and sand that the creek had deposited in its bed were a few kernels of grain and some gleaming yellow pebbles of a mineral that, although useless for any practical purpose, came to be seen as extremely valuable.

With the coming of spring, the sun started to warm up the steep south-facing side of the gully, while the deep-frozen bison at the bottom of the ravine remained in the shade. The dark, frozen ground absorbed the heat and began to thaw. By night it froze again, but only on the surface. Day by day, the layer of thawing earth grew deeper, from centimeters to inches; and there was no possibility of draining it, for the frozen ground beneath remained impermeable.

Finally, one sunny afternoon, the thing happened. The waterlogged mass of earth that had so long been poised on the incline, perhaps triggered by an earthquake, slid down with a crash and buried the half-eaten carcass. And down at the bottom of the gully, in the shade, the ice-cold water froze anew. And so that bison was preserved in nature's own refrigerator: it was now frozen into the permafrost.

When the body froze, all microbial activity came to a standstill; so the meat remained fresh, except for part of the abdomen that had gone bad before the freezing process began. Now, during centuries and millennia, the dormant microorganisms died. Everything became sterile, with an asepsis that not even the most modern hospital could hope to emulate. Additional solifluction earth piled up, burying the body ever more deeply. It was now sealed into the permafrost of Alaska, the timeless *tiaele* extending hundreds of feet down.

When the bison died, Europe was still Neandertal country. But the earth spins around the sun; year is added to year, century to century, millennium to millennium. Hu-

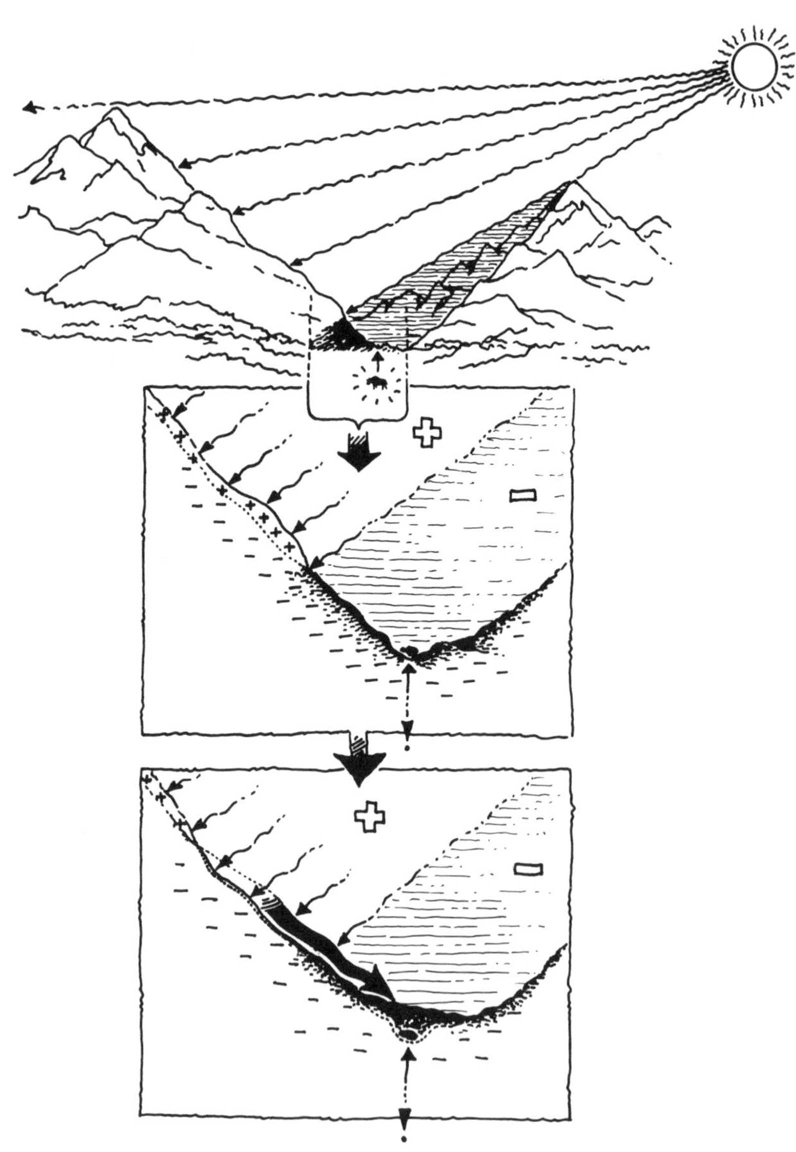

man beings of modern type enter Europe. Neandertal dies out. Magdalenian people hunt the steppe bison and render its image in engravings and paintings—at Lascaux, at Altamira, at Niaux, at La Madeleine. The great land ice begins to melt and shrink. The last steppe bison dies and is succeeded by the modern wisent; the elder species vanishes from human memory. The first civilizations grow up. The Sumerians invent cuneiform. The pyramids are built. The Greeks and the Romans have their time. Charlemagne keeps court in Aachen. Leif Eriksson, then Columbus cross the Atlantic. The Europeans conquer America. Vitus Bering dies at the border of ancient Beringia, now half-drowned by the rising seas. The steam engine ushers in the modern age. Man goes to the moon. But for the dead bison, time stopped ages ago; all processes were suspended; the bison rests in a changeless present, as fresh as when it froze a thousand generations ago. Just as fresh, its skin and flesh torn by lion teeth and raven beaks, as if it had happened a few hours ago. Yet its own species is now extinct since more than 10,000 years, and no human being has eaten the flesh of *Bison priscus* since the time of the Magdalenian hunters.

Until now.

This is a reconstruction of what probably happened. Even in the permafrost, the finds are usually bones without soft parts and this shows that several improbable, but not impossible, conditions have to be fulfilled to get a frozen carcass.

Let's return to the present. Gold prospectors use powerful jets of water to thaw and rinse away the soil and reach the old strata where gold can be found. In the middle of all this they often find the bones and teeth of Ice Age animals. Professor R. Dale Guthrie, a paleontologist at the

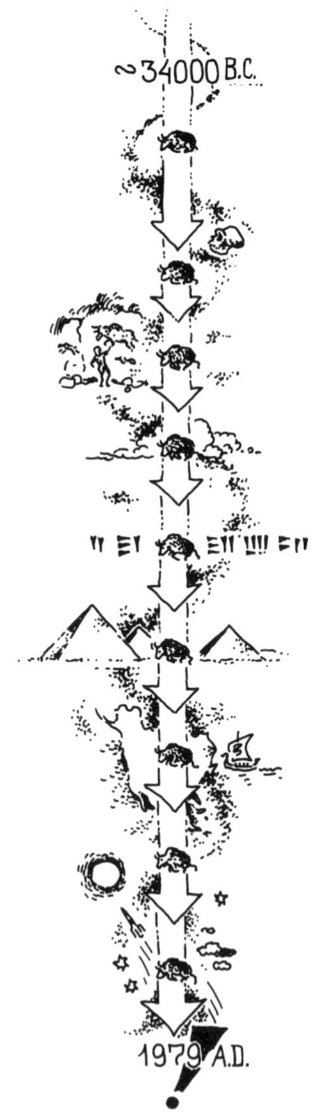
~34000 B.C.
1979 A.D.

University of Fairbanks, is an old hand at the trade, and when in the summer of 1979 he heard that something that looked like an animal foot was sticking out of a layer of earth, he immediately suspected an important find. The carcass was quickly salvaged by Dale and his wife, Mary Lee, and was placed temporarily in cold storage.

In the spring term of 1980 Dale spent part of his sabbatical at the Geological Institute in Helsinki and made the acquaintance of Eirik Grandqvist, then chief curator of the Zoological Museum. It soon became clear that Eirik was the right one to preserve and mount the Ice Age bison for the museum at Fairbanks, one reason being that taxi-

The 36,000-year-old steppe bison "Blue Baby"—mounted by Eirik Granqvist in the University Museum at Fairbanks, Alaska. Killed by a lion, the bison is shown at the moment of death. The hair is loose but the skin is well preserved, with the exception of parts along its back that were eaten by its attacker. Portions of the frozen meat were well preserved and tasted fine.

dermy is a dying art in America, at least for big mammals. The work was finished in the spring of 1984, and Eirik himself considers it the high point of his career. The bison may now be viewed in its death position, resting on its belly, its head with the imposing horns lowered. The skin is well preserved, although most of the hair has come loose. The skin and the other exposed soft parts are covered in places by vivianite, or blue iron earth, which is formed by phosphorus in the tissues reacting with iron in the matrix. This earned the fossil the nickname "Blue Baby." It is the third preserved Ice Age animal in the world to be mounted, and the first in America.

The meat in its abdomen had spoiled before the bison was completely frozen. But in the neck area small pieces of meat were found attached to the skull. The lions had left so little there that it had frozen through while the meat was still fresh. When it thawed it gave off an unmistakable beef aroma, not unpleasantly mixed with a faint smell of the earth in which it was found, with a touch of mushroom. About a dozen of us gathered in the Guthries' home on April 6, 1984, to partake of *Bison priscus* stew. The taste was delicious, and none of us suffered any ill effects from the meal—which meant that a number of organic molecules were returned to the biosphere after a thousand generations!

The Swedish prehistorian and writer Axel Klinckowström once made the following point. Imagine a dinner table set for a thousand guests, in which each man is sitting between his own father and his own son. (We might just as well imagine a women's banquet, a thought which, typically enough, didn't occur to Klinckowström.) Then, Klinckowström notes, at one end of the table there might be a French Nobel laureate in a white tie and tails and with the Legion of Honor on his breast, and at the

other end a Cro-Magnon man dressed in animal skins and with a necklace of cave-bear teeth. Yet every one would be able to converse with his neighbors to his left and right, who would be either his father or his son. So the distance from then to now is not really very great.

And certainly we, who partook of the *Bison priscus* banquet, had the momentary dizzying feeling that, for a moment, a bridge was spanning the ages.

AFTERWORD

A few of the chapters in this book have appeared in Swedish in earlier versions. The chapters "Why Should We Feel the Countryside Is Beautiful?" and "Thinking Biologically" appeared in *Nya Argus* in 1985 and 1986. "Biology and the Bookshelf" is based on a talk delivered at the international writers' meeting at Lahtis in 1983. But they were written with the aim of including them in this book.

For their help and good advice in various connections I wish to thank Olli Alho, Mikael Fortelius, Dale Guthrie, and Jakob af Hällström. The collaboration between myself and Viking Nyström, who illustrated this book, as always was pleasant and inspiring.

INDEX